Advance praise for *Futureproofing Humanity*:

In *Apocalyptic AI*, Robert Geraci was far ahead of most of us in calling out the foibles of a small group of technologists who were attempting to foist off digital technology as a new God. In his new book, *Futureproofing Humanity*, he poses the same important questions about humanity, computing, and religion. Like Stewart Brand before him, he points out that the power of these new computing and genetic engineering technologies brings new responsibilities for humanity, and if we are to survive as a species we must take that responsibility seriously.

— John Markoff
Author of *Whole Earth: The Many Lives of Stewart Brand* and *Machines of Loving Grace: The Quest for Common Ground between Humans and Robots*

———

An insightful perspective on recent technology developments, the potential impact on our survival and our civilization, and on the possible futures envisioned by some of our contemporaries. Informed by a surprisingly rich history of thinkers worried about the possibilities we face today, *Futureproofing Humanity* is recommended to anybody concerned about the survival of humankind.

— Matthew T. Mason
Professor Emeritus, Carnegie Mellon University Robotics Institute and Chief Scientist, Berkshire Grey

———

In *Futureproofing Humanity*, Robert Geraci offers a timely and important book. Part intellectual history, part cultural critique, and part fiction—*Futureproofing Humanity* examines how genetic engineering, artificial intelligence, and space exploration have become the vessels for humanity's oldest religious longings: immortality, resurrection, and cosmic purpose.

What makes this book so valuable is its refusal to simply debunk or celebrate the transhumanist vision. Instead, it takes seriously the religious dimensions of technological utopianism while asking hard questions about who gets saved and who gets left behind. From Nikolai Fedorov's "Common Task" to Ray Kurzweil's Singularity, from cryonics to colonizing Mars, the author traces how the promises once made by gods have been transferred to silicon and code—and what we lose and gain in that translation.

This is essential reading for anyone trying to make sense of our current technological moment, when billionaires promise us the stars while the planet burns and AI chatbots pass the Turing test. The author writes with clarity, wit, and a scholar's rigor, but also with genuine care for the human stakes involved. He reminds us that civilization—with all its flaws—is worth preserving, and that the myths we tell about our future shape the choices we make today.

— Ilia Delio, OSF, Ph. D.
Josephine C. Connelly Endowed Chair Villanova University

FUTUREPROOFING
HUMANITY

*Existential Risk and the Technomyths
of Human Engineering, Artificial Intelligence,
and Our Future among the Stars*

ROBERT M GERACI

CONTENTS

This book is dedicated to my sons, Zion and Dorian,
who survived (perhaps enjoyed?) homeschooling with me, and who
have changed me for the better in ways beyond count.
I hope they always feel the love of learning, the power of curiosity,
the desire to see the world as it is and as it can be.

A NOTE ON THE
NATURE OF THIS BOOK

This book is a departure from my usual work in two aspects: I've parted ways with traditional publishers and I've forsaken all the carefully (and productively) indoctrinated modes of citation that characterize all my previous publications. Both of these were deliberate choices, risks that I hope will pay off.

Regarding the second, the book does mention lots of authors and thinkers along the way, and it includes a long bibliography at the end. Some 300+ sources. But in the interest of making the book optimally readable for a wide audience, I have largely ignored the scholarly convention of citing every idea in every location. All I can say is that this is deliberate. It is not an attack on that kind of scholarly precision and in no way means that this is the future of my writing. It is only that I wanted to try something different, something that might more easily intersect with the interests of an increasingly choosy, and perhaps diminishing, number of book readers. I hope that readers will see me as contributing to a millennia long tradition of mythmaking, and not simply a centuries long tradition of scholarship. There is rigor in both, and I have largely switched the one kind for the other.

As to the first departure, I grow more disenchanted with my academic publishers with every year that passes. I am, of course, appreciative to them for bringing my books to market. Oxford University Press has been particularly good to me in publishing three of my books. But no amount of conversation about their pricing model has shifted them from the absurd expectation they should sell exclusively to libraries. No one will pay $50 for a Kindle book (which is common in my books' strangely varying prices). To my mind, selling 50 at $1 is a better business model than zero at $50. But my pleas fell on deaf ears. One publisher even refused to sell me the digital rights to a book that sold only one Kindle copy (to me). Their model is broken, and nothing shows that more clearly than when a book like *Apocalyptic AI* long ago recouped its costs, paid the publisher, paid me royalties (and still does), and yet gets priced to ensure no one will buy the digital version. The Kindle version can be offered at no marginal cost but generally lists between $35 and $60 (why is it so expensive? why does it change price? who can say?). So, I'm trying to self-publish this book. I have no idea if anyone will find it, but at least I can guarantee a price below $20.

One way in which this book continues my standard approach to work: no generative AI was used anywhere in its research and composition. If I thought a large language model (LLM) could replace my thinking, I'd just give up on the writing. I hope that readers will continue to treasure the human labor that requires a start to finish composition of an idea or set of ideas. This is not to say that there are no useful purposes to LLMs; it's just that they are few and far between, and they don't include the authorship of scholarly books and/or modern mythologies.

PROLOGUE
A FUTURE FAIRYTALE

Once upon a time, people had faith. Their gods and spirits were mercurial, generally uninterested in what the people wanted or what they'd done. Most often, the people just wanted to be left alone, to live far from the dangerous glance of the divine. The people asked for little more than health and safety in this life, and then, perhaps, a better life beyond. Although the gods and spirits said little (and demanded much), the people were moderately certain that they guided the world and offered shelter. The gods and spirits punished, and the people suffered, but better days were promised. The people lit fires and gave gifts. They crawled in the ashen dust and they climbed mighty peaks. They sought assurances in a difficult world. And maybe they even got what they asked for. Sometimes.

Over many years, the people designed their own protection, found other ways to guide the world. They gathered together and they learned. They grew more food, and better. Their buildings stood firm against wind and storm. They had so much of everything that they couldn't possibly use it all. Eventually, it seemed that they, themselves, were in charge. They dreamed and they built. They delved into the secrets of the world's foundation. Their tools, once made of bone and flint, were now steel and silicon. A spark of power flowed in them.

The electric spirit gave movement to the world, to their tools. These machines took on life, acting of their own accord, and helping the people. The machines produced marvels, made life easy. And so came the day when many of the people no longer needed their old gods and spirts at all. But they needed the machines, because the machines were built to keep people safe and to help people be wise.

They looked upon the world, the world of their machines, and wondered.

———

Wonder recognizes awe and fascination, and also the unknown. It is a word of both fear and attraction. And as human beings wonder at the world we have built, we are caught by that dynamic. Our machines enthrall us, drawing millions of YouTube views when robots dance. But they also terrify us, snatching from us our traditional jobs in the factory or office and creative jobs in art.

The mixture of attraction and fear is at the heart of religious life also. In the early 20th century, the Christian theologian Rudolph Otto spoke of his god's *fascinans* and *mysterium tremendum*. The first is the lure of care, purpose, and salvation. The second is the terror of distance, damnation, and the god's "wholly other" nature. Otto saw our human response to the divine as a convergence of these opposites. Many years ago, I published an essay about how we see robots in a similar way; or, at any rate, how we write about robots in that way. Our science fiction tales of robots cleave to our stories of the divine. That essay was before driverless cars roamed the streets of major cities and even before they raced across the desert (in pursuit of public acclaim and military dollars). It was based on science fiction, not actual science.

But now our technologies are catching up to the stories. Not just robotics and AI, though perhaps chiefly these: modern technology promises both salvation and damnation. Many people hope for a life of leisure and universal basic income while others fear that humanity will soon go extinct. Otto may have welcomed the twin experience of *fascinans* and *mysterium tremendum*; but faced with this technological crossroads, human beings cannot be complacent. We cannot simply accept the possibility of damnation alongside salvation. Human beings must look to the future and preserve it. We must futureproof humanity.

To futureproof something is to ensure it never becomes obsolete. That is, if you want a device or a system to be useful well into the future, you have to make sure it continues to function and it will still have a purpose despite changes to its environment. Prior to the 20th century, humanity (as a whole) didn't have to worry about whether there would be a place for human beings in the future. But our military capacities, our inadvertent changes to the environment, and our awareness of geological and cosmic forces beyond our control have changed that. Futureproofing humanity has become a priority.

Futureproofing is built on the transfer of faith from gods and spirits into technology, and this book describes how we've come to rely on a few key technologies and place our hope for the future within them. It is thus a sort of fairytale, suggesting that there are powers beyond mortal ken and a hero's journey to finish. Fairy tales are not just children's literature and they are not bound by the confines of fiction. We cannot know how many or what kinds of meanings that the authors or hearers of ancient (or even more recent) fairy tales might have intended or seen. To know whether the tales were intended or understood as literal in addition to metaphorical is beyond the archeology of knowledge.

But modern fairytales are frequently told as literal, and such is the case of futureproofing humanity. The stories are not just science fiction, they are literal science. They appear in pop science books and even in research grant applications. The stories of our future, of whether we can survive our own genius, our genetic manipulation of species and our ability to bring life to inanimate matter, are stories intended as literal. But they are too much like Rudolph Otto's divine and not enough like a great epic.

And so to respond, to join their conversation, I offer a new fairy-tale. This is a story that interrogates the promises made by our technological elite. It is not an explanation of technology, but rather an exploration of such explanations. And in that search, it offers a vision of civilization and technology to counter the loud narratives in global culture. The narrative that declares all human labor should become AI labor. The narrative that says our technologies matter more than we. The narrative that promises we must relentlessly pursue profits as swiftly as possible. All those stories matter but I wish to offer a new path. The world is, in fact, fraught with dangers. Some of those dangers are of our making and some are not. Our best chance at a long human history is to invent our way forward, to preserve human civilization through the tools that created it. In drawing on our technological promises, most especially our faith that human destiny lies among the stars, I weave together the futurist promises of our tech elite in the hope that we can build a better world.

This book is, however, the kind of fairytale that could become a ghost story. There is no guarantee of a happily-ever-after, though such hopes are at the heart of the book. Today's promises about the future are haunted to the core. A 19th century philosopher argued that every thesis includes its antithesis, every argument its own counterargument, and in that way the promises of technology include their own potential undoing. So this book makes no assurances of a happy outcome; but it supposes there may, indeed, be enchantment lurking around the corner. There are certainly glimpses of a happy ending to come, but our doubts and fears permeate them. We can have hope, but we wonder if we ought to.

Twenty-first century enchantment is firmly ensconced in technology. I'm not the first academic to have written on this subject, though over the past twenty years I think I've made some unique contributions to it. One of my starting assumptions has always been that human beings are relentlessly interested in the promises of myth, folktale, and religion. We want a world full of purpose, flush with

4

meaning and magic. And so even people who have rejected every traditional religion in the world find themselves bending in the arc of unfulfilled purposes, resurrection of the dead, personal salvation, and cosmic redemption. Even when we ignore Jesus or the Four Noble Truths or the influence of our ancestors, we still look for the kinds of things those people and ideas used to offer.

I grew up at a time when you knew the local fallout shelter and everyone saw at least one ridiculous video of what to do in case of nuclear attack. It was all fairly terrible advice, but I suppose that it gave us a sense of agency, a sense there was something to be done. The threat of nuclear war remains in the 21st century; and alongside it looms an array of dangers so awful to contemplate that many of us get ready to "duck and cover" once again. But there may be real opportunities to shelter ourselves from the dangers lurking all around us. We can, perhaps, invent our way past the threat of mutual assured destruction. In this era of robot taxis on Earth and rovers on Mars, maybe we can imagine and build our way to a safer, more secure world.

The future of humanity will be a story of biological evolution, technological progress, and – perhaps – immortal salvation. Not the heaven of Christianity or the Nirvana of Buddhism, but a glorious future through technology. In the pages to come, I describe how genetic engineering, robotics, artificial intelligence, urban design, and spaceflight have become infused with our hopes for immortality, resurrection of the dead, and other religious dreams. This is a book about why that is happening and, to some extent, provides a justification for it. We are, after all, faced with enormous dangers: a Bermuda triangle of climate change, robotic automation, and political conflict. How are we, as a species, to emerge unscathed after sailing into such a torrent of risks?

"Once upon a time" is how fairy stories begin, and once upon a time it would have been possible for entire societies to believe that divine guidance would help humanity navigate its way into the future. Gods, ancestors, or spirits will keep us safe. Although I am firmly religious, I do not see good reason to believe that's the case. We don't know how many people of the past *actually* felt confidence in those religious promises; but we know the promises were widespread. Traditional religious commitments remain common today; but, as commentators have noted since the time of Nietzsche and Freud, people now express their doubts publicly (as I did just a few sentences ago). Confidence has been shaken.

In a world where doubt and uncertainty prevail, technology has become the locus of our deepest desires: the search for immortality and cosmic purpose taken out of the hands of gods, spirits, and ancestors. This shift from traditional religion to technical religion is probably justified. The risks we see are real risks. The human species is in danger. Most of the dangers we face are those of our own creation. Some are not. Regardless, if we are to survive another century and the centuries after that, we need to find a path forward. Perhaps the fervent believers in futureproofing humanity are right to offer us hope.

Because they offer resurrection, immortality, and cosmic meaning, the promises of futureproofing are part of a religious story we have been telling for millennia. In one way, shape, or form, just about every religion offers a reason to hope for better days to come. The mythos of futureproofing coalesced out of those ideas, combined with new scientific and technological insights. It is the translation of old dreams of salvation into a technological worldview of genetic engineering, artificial intelligence, urban design, and – bringing them all together – spaceflight.

Humanity first caught a serious glimpse of these technological possibilities in the late 19th century, and that is when we begin our story. The Russian Cosmists and the western transhumanists believed that we would one day use science and technology to achieve transcendence: to resurrect the dead and acquire immortality. While their technologies were barely sufficient to imagine these possibilities, let alone accomplish them, they prompted increasing faith and increasing innovation. Out on the fringes of human imagination, an alliance of scientists, futurists, and science-fiction authors contemplated how technology could satisfy our keenest desires.

That fringe has become mainstream. There is a triumphalist narrative that seems to think such mainstreaming is inevitable, some kind of unfolding historical necessity; but lots of fringe science never became popular. For example, very few of us think we're living on the inside of a sphere. In this book, I discuss how the technological salvation of humanity has, in fact, become a dominant narrative. Words like cryonics, cyborg, and artificial intelligence emerged in the 1950s and 1960s. The word robot traces to the 1920s. When those words arose, few but their creators saw them as The Future. But now, with fear or hope, people across the world do exactly that. I will engage with those who fear the futureproofing technologies near the end of the book; but most of our journey will be with the dreamers and optimists, those who believe that technology is a necessary and vital part of human salvation.

The hopeful vision of futureproofing resists a terrible fear: that our time on Earth is short. It is typical to complain that people age too fast and die too young, and those concerns are central to many religious worldviews. But the religion of futureproofing has a new fear; or, rather, a fear that has been made far more pressing than it was in the past. In the 21st century, we have no choice but to fear for the survival of our entire species.

That kind of fear, now labeled "existential risk," seems both eminently reasonable and emotionally overwhelming. For the most part, humanity hasn't had to wonder whether there would still be human beings in five, ten, or twenty years. But those fears are on the rise, and with cause. We cannot predict the outcomes of climate change, though we know it will make life unbearably hard in many or even all parts of the world. We know that some of the asteroids hurtling through space are large enough to wipe out all life on Earth. We rush joyfully to replace ourselves – as students, as employees, even as companions – with artificial intelligence, and we wonder if that's a prelude to a more permanent replacement. Perhaps Skynet is more than just science fiction? And meanwhile, thanks to the looming threats of computer surveillance, political authoritarianism, and global war, the *Bulletin of Atomic Scientists* keeps moving the hands of its Doomsday Clock closer to midnight.

Before anyone bandied about the phrase existential risk, roboticist Hans Moravec dreamed of a future where we could forever stave off extinction. In his 1988 book, *Mind Children*, Moravec preempts the existential risk crowd: "the universe is one random event after another. Sooner or later an unstoppable virus deadly to humans will evolve, or a major asteroid will collide with the earth, or the sun will expand, or we will be invaded from the stars, or a black hole will swallow the galaxy." Within a sufficiently long view, the extinction of life on Earth is inevitable. But he also proposes that "by growing rapidly enough, a culture has a finite chance of surviving forever" and he happily imagines how something could "restructure itself so as to function indefinitely even as its universe ended." As in so many things, Moravec preempted the conversations that now govern our culture's engagement with technology and the ultimate fate of our species.

We do not know what the future holds, but we know it comes with risks. And those risks accelerate the new technological religion. Futureproofing humanity holds that technological progress will offset or even overcome the risks we face. This view of the future is about making sure that human beings are proofed against all the risks that the future holds. Those risks are individual and they are collective. The futureproofing religion says that the technologies bringing personal salvation will also save our species. We will gain superpowers, resurrect the dead, achieve immortality, and explore the universe. In reaching for the stars, humanity will become divine.

Those transcendent aspirations, divorced from traditional religions, point toward the new technological vision. Religion, according to David Chidester, is "the negotiation of what it means to be human with respect to the superhuman and the subhuman." Religion is how we find meaning in the world, hold beliefs, participate in rituals, create art and literature, maintain doctrines, build communities, and more. When those kinds of things entail angels, demons, ancestors, gods, or cosmic consciousness, they become religion. In the case of futureproofing, the risks are mundane but the solutions are not. We may be on the road to superhumanity.

Futureproofing humanity means to proof us against the future, to ensure our survival against anything that might come our way. In the 21st century, futureproofing is a fairytale, a myth, and a religion, and, perhaps, our only hope.

CHAPTER ONE
GHOSTS OF FUTURES PAST

In the 19th century, scientists relocated humanity to a new place in a long and uncertain cosmic history. And now that uncertainty prompts us to move fast or else risk our future. Two scientific classics, in particular, rewrote human history, focusing on the laws of nature and implying potential catastrophes for our world and our species: Charles Darwin's *On the Origin of Species* (1859) and Charles Lyell's *Principles of Geology* (1830-33). Because these books dealt in origins, they were always also about ends. If Earth, the solar system, and humanity had natural beginnings, they will have natural ends. The destruction of the Earth or the extinction of humanity are only implied in such a history; but that implication cannot be ignored. It is the foundation that supports contemporary futurism; it motivates 20th and 21st century scientists and philosophers to pursue a technologically transcendent humanity.

Before the naturalistic view of human evolution and extinction emerged, medieval and early modern Europe saw many Christian futurists predict the imminent end to the entire world. But they never stopped believing that they – god's elect – would be saved. By the time humanity had come to see the cosmos as an unimaginable gulf and to recognize the momentary presence of humanity within that, we learned a fear that our species could, per the saying, go the way of the dinosaurs. There may be no salvation, no final victory.

Twentieth century thinkers took the Darwin/Lyell worldview but argued that the very forces enabling us to understand the cosmos (science and technology) would guide us past the risks of extinction. They hung onto their faith in a universe aimed at their own salvation. Other species might be subject to cosmic luck or the ineluctable forces of nature, but humanity would rise above. Humanity's inheritance would be immortality, promised not by gods but by scientific ingenuity.

The story of science and of humanity's technological mastery is a long one, and this book makes no pretense toward being comprehensive. So rather than start "in the beginning," we will launch our understanding of today's fears, hopes, and religion of technology after the Scientific Revolution, after the Enlightenment, and midway through the Industrial Revolution. There will be necessary moments for reflecting on earlier times, but we will begin the story of existential risk and scientific salvation in the 19th century. It was a crucial time for our understanding of the world and our presence in it, and thus crucial to everything happening in the 21st century view of humanity and our increasingly tenuous future. We will have an opportunity to think outside the western cultural box, but the story of existential risk and salvation through technology is primarily a western one.

It's not obvious, of course, why risk, religion, and technology would be part of any one story at all. Many people continue to believe that science and religion are at war, in which case it hardly seems reasonable that we would build a religion out of scientific progress. But the conflict narrative is wildly overblown, largely invented in the 19th century after Lyell and Darwin met with a touch of resistance here and there. Inaccurate histories accused religious people of believing absurdities, like the flat Earth theory. In fact, almost no one believed the world was flat. Why would Columbus have set out on his voyage of conquest and proto-capitalism if he thought such a thing?

More perniciously, many people think the Trial of Galileo proves religion and science to be at war. The ongoing certainty in this leads many people astray. But, in fact, Galileo was a devout Catholic, a position unquestioned even by his most ardent opponents. Their conflicts were political and scientific, not really a debate between theology and natural science. Certainly, there have been times when religious people angrily rejected scientific theories (though rarely technological results). But conflicts often have entirely different explanations at their core, and there are at least as many historical examples in which religious motivations underpin scientific exploration.

The religious incentive to scientific discovery is one that is both ancient and modern. I have had scientists directly tell me that while they do not believe in the literal claims of ancient myths, the stories provide inspiration toward technological development. For example, if ancient sages in Indian folklore could speak telepathically, how might we build a technological platform to do so in reality? The flip side of that is contemporary religious figures like Paramahansa Yogananda, well-known for sharing Hindu beliefs and practices across the United States in the mid-20th century. He says in *Autobiography of a Yogi* that his teacher could be in two places at one time and explicitly uses radio and television technology to "prove" that such things are possible: "in its own way, physical science is affirming the validity of laws discovered by yogis through mental science." While I differ with the sage on this question, it's worth seeing how many people around the world see direct linkage between religious claims and scientific claims.

Absurd claims that connect religion and science are not the exclusive territory of any particular religious community. Just two weeks before I wrote these words, I had someone tell me that it is "scientifically proven that Muhammad cracked the Moon." Near the end of 2025, well-known entrepreneur Peter Thiel dubiously asserted that the antichrist was out there, planning to use AI legislation to usher in the end of the world (he literally named environmental activist Greta Thunberg and AI safety advocate Eliezer Yudkowski as the sort of person looming with this evil agenda).

The odd conjunctions of religion and science are just one strand in a thread of interconnections that have long held religion, science, and technology entwined – other interactions are more rational. In fact, some scholars have even suggested that science emerged out of religion. In *The Elementary Forms of Religious Life*, Emile Durkheim argues that it is through religion that we learn to make connections between seemingly unrelated facts. Right or wrong, he believed that ability was at the root of the scientific method. In any case, the cultural context of science regularly draws on religious ideas, practices, and institutions. As Catherine Newell argues, "science *needs* religion – the historical rituals, the patterns of faith, modes of personal belief, and habits of the heart that define both institutional religion and private spirituality are also often the root of scientific endeavor." Religion helps drive science even if its assumptions, methods, and conclusions are far separated from laboratory work.

It is this political process that ensures there will never be a real divorce between science and religion. And that political process also drives technology. In *America as Second Creation*, David Nye shows how religious motivations infused the technologies of American expansion and conquest in North America. The axe, the railroad, the telegraph. These technologies and more were the signs of a divine mandate, the Christian obligation to conquer and control.

While the impact of religion was explicit in many technological contexts – from the pursuit of Adam's supposed perfection and power to the belief that ancient Indians possessed flying machines – the intersections of religion, science, and technology have been subtler in recent decades. This subtlety results from the rise of secularism. Roughly speaking, secularism refers to efforts at divorcing religion from politics, public life, and, of course, science. There are still members of religious communities that want to force everyone else to follow their rules; but for the most part secularism implies that a specific set of religious rules and beliefs should apply only to those who commit to them willingly.

The power of secularism is not the erasure of religious belief and practice, however, but rather their submersion into other forms of culture. We find transcendence in abstract art, utopian dreams in architecture, and communion with the faithful in March Madness. We seek spiritual peace in decluttering our homes and escape the mundane through the enchantment of videogames. We have found so many different ways to seek religious satisfaction in everyday life that it has become genuinely commonplace. Part of that process has been to relocate traditionally religious goals into our scientific world-view and technological devices.

Consider the pilgrims who attended the opening of every new Apple store early in the twenty-first century. They had a vision, brought to them by their saint in Silicon Valley. They found meaning in their consumer religion, and in their community of fellow believers. The religion of Apple prompted more than one book and a legion of popular and academic essays.

The sanctity of Apple and its iPhone follows in the footsteps of early modern Europeans, for whom science and technology were religious pursuits. The mechanical inventions of the 12th and 13th century were as likely to be built in a Catholic monastery as in a merchant's workshop. And over time the things *sought* by monks – redemption of the world and its inhabitants – came to infuse the things *built* by monks. Nearly all of Christian Europe's educated elite leaned into the idea that science and technology advanced in the service of their god. As public commitment to their god waned in the 20th century, science and technology remained as the sole guarantors of religious redemption.

It's possible that redemption grew increasingly secular because the timeline of the world shrunk our sense of human significance. We have learned that the history of the cosmos is measured in billions of years, a process that beggars the human imagination. The Earth is more than four billion years old and life appears to be around three billion years old. And yet human history is just the last few thousand years. Throw in human prehistory and some human-like ancestors and you get only a few hundred thousand years at best. Our presence on the Earth is exceedingly recent! We weren't here at the beginning and – if the subjects of this book are to be believed – we may not be here in the end. Humanity sits within that cosmic span and our imagination of the universe has expanded from one solar system into countless galaxies. We are a very recent addition to one planet in one solar system in one galaxy in one immense universe of space and time. But locating ourselves within that cosmos began when we started seeing our own planet more clearly.

The history of geology is not winning beauty contests in the public sphere, but it deserves recognition as we trace the present and future of humanity. In particular, a turning point came in the 1830s when Charles Lyell published his three volume *Principles of Geology*. Along with their unrepentantly empirical approach to earthly phenomena, these volumes presented a radical but irreplaceable new faith position: uniformitarianism. This is the unprovable belief that the laws governing natural events are uniform across space and time. I cannot *prove* that light moves according to the same principles on Earth as it does in the Andromeda Galaxy; but if I do not *assume* that, then I cannot learn anything about the Andromeda Galaxy. It is our presumption that the cosmic rules are the same everywhere and everywhen that allows science to happen at all. So, Lyell's rules of engagement are absolutely vital even if we cannot guarantee their truth. First, we came to realize that the way the world works right now helps us understand its history and properties (according to geologist James Hutton's maxim, "the present is the key to the past"). Then we started to understand the planet's history. And then we started to see ourselves as rather small fish in the geological pond.

The uniformity of natural processes across cosmic history includes the development of humanity: our species' evolution via natural selection (along with genetic drift and sexual selection). This recognition accelerated the secular commitment of humanity. As many have noted previously, the seeming miracle of human life got upset when we realized that many species had emerged and had gone extinct over the past few billion years. We started developing alternatives to what has been lately called "the god hypothesis." We had less need for traditional religions that situated humanity in the cosmos because our presence could be explained through the long, slow march of evolution. This is not to say there is no god behind the evolutionary process; it is quite beyond me to judge that question. Gods or not, opportunities arose to rethink our place in the universe.

I am not suggesting that people stopped wanting to be religious. We never stopped looking for meaning and enchantment in the world. We never stopped hoping for redemption. But our clearer vision of humanity's place on Earth made explicitly religious redemption less likely *regardless of whether or not any of our gods existed*. If we're less special in the universe, we might be less special to the gods. And the gods might not be real anyway.

There have been dozens of influential thinkers to suggest that human beings invented our gods for one reason or another, and that those gods are illusory. Some of these theories are better than others, or have more use than others when we attempt to explain the world around us. None of them prove that the gods aren't real. This is emphatically not the book to debate whether they are. I am instead describing what has happened in our social world, where we often lack conviction in the old promises of religion but have every bit as much interest in them.

Among atheist explanations for the existence of religion, the exchange model of William Bainbridge and Rodney Stark is one of the best. They argue that human beings are economic at heart: we constantly look to divvy up labor and trade for things we don't have. So I learn to bake apple pies and then trade them to someone else who is excellent at hunting turkeys. If there's a thing that I want, I go looking for a trading partner. But Stark and Bainbridge rightly note that there are no trading partners for certain things that human beings want: a meaningful life, perfect health, eternal youth, the resurrection of our loved ones, etc. And so they suggest that human beings, out of desperate want for those things, invented the trading partners. In essence, Stark and Bainbridge argue that we want things like immortality so badly that we invented someone(s) who could give them to us. I don't know whether that accurately explains ancient history, but it turns out to be an outstanding explanation for how science and technology came to promise religious salvation in recent history.

Suppose there are people who believe that there are gods who can provide them with meaning, purpose, happiness, and immortality. They happily engage in the trading practices demanded by their half of the transaction: sacrifice, moral behavior, and ritual observance. They hope for meaning, happiness, and immortality. But slowly, they find that their sense of meaning is eroding. Perhaps this is because they realize what very small fish they are, and in what a very large pond. They feel insignificant, and thus meaningless. It doesn't matter whether or not the gods are real. What matters is how people feel, and now they feel doubt. Given the shock to their system, they wonder about all those transactional promises. After all, no human being is always happy. So perhaps immortality is lost to them as well. Into this sad world of emotional crisis steps a technological innovator, one who argues that there is, in fact, cosmic purpose and that ultimately humanity will indeed achieve perfect happiness and even immortality.

The new religion of futureproofing proposes that we will become gods ourselves; so gods don't exist *yet*, but they will. At least some of the existential threats are inevitable. For example, there will be another supervolcano explosion at some point. There will be another large meteor that hits Earth. They may be millions of years away, but these accidents are coming. Alternately, if somehow these things don't come around, the Sun will reach the end of its lifespan a long time from now, and its life-stage transition into a red dwarf will be the Earth's demise. At some point, the Earth and its inhabitants will die. We seek redemption from this and a better world to come. It appears that opportunities to deliberately evolve our species will be key to the continued survival of human consciousness, and technology enthusiasts have built an industry out of telling us that our journey is just beginning.

That transition is what this book is about, even while it is also about the haunting prospect that those futurists might be wrong: that their fear for the future is warranted but their promise of salvation is not.

If the presence of humanity is recent and the result of geological and biological accidents (and maybe even the occasional touch of divine intervention – who knows?), then it may be rational or even necessary that we fear for the future. Other species went extinct as the world spun inexorably toward the Anthropocene. For one reason or another, perhaps we are at risk of extinction also.

A variety of contemporary thinkers define human extinction-level events as "existential risks." An existential risk is anything that could legitimately result in the demise of our entire species. In the world of natural accidents, collision with an enormous asteroid or the explosion of a supervolcano could lead to mass extinction. But human beings could also ensure our species' demise through nuclear war, genetically engineering a supervirus, or simply burning so many fossil fuels that the environment collapses under the combined influence of natural disasters, rising temperatures, shifting patterns of wind and wave, and the extinction of key species, such as pollinating insects.

Fear of those existential risks now drives commitment to the new technological religion. That new religion has no need of gods. After all, the gods might themselves be existential risks ... just ask Noah from the Bible, or his predecessor about whom the story was first told, ancient Mesopotamia's Atrahasis (known as Utnapishtim in the *Epic of Gilgamesh*). Instead of looking for divine redemption, the new religion seeks salvation in technological savvy and places its faith in unstoppable scientific development.

Existential risk, redemption from it, and the transcendence of human limits are part of a collective package, a set of interdependent ideas that no longer exist separately. The connection between human transcendence, technological progress, and future redemption emerged long ago, became especially salient in the 20th century, and now anchors an entire worldview. That perspective runs throughout the entire futureproofing agenda described in this book; it's called transhumanism.

A brief history of transhumanism

Transhumanism is the philosophical or religious commitment to transcending human limits through science and technology. In a meaningful sense, all human beings are transhumanists: we use glasses to see, clothes to stay warm, mobile phones to connect with people, and airplanes to fly. We are all in the constant process of using technology to overcome our limits. But there is a generally recognized (if often blurry) distinction between that everyday aspect of human life and a commitment to using technology to perfect the body, become immortal, or resurrect the dead. The Christian theologian Phillip Hefner once referred to this as the difference between "transhumanism" and "Transhumanism," though his typographical distinction didn't catch on.

The emergence of capital-T versions of transhumanism was a gradual one. Google's ngram viewer indicates that the first published usage of the term was in 1951, but as of 1995 the concept appears to have entered an exponential growth curve, rising in frequency at meteoric rates (even if it remains uncommon). The term owes its existence to biologist, conservationist, and cultural ambassador Julian Huxley, who coined it in a 1951 essay published in the journal *Psychiatry*. Huxley retained the term, more famously using it in his 1957 book, *New Bottles for New Wine*. Perhaps unsurprisingly, it was at the time the Internet started to become mainstream that transhumanism began percolating through pop culture.

Google Books ngram report on the published frequency of the word "transhumanism" for the years 1800-2019

Though a word for it did not exist until mid-century, transhumanism emerged decades prior (Huxley used the phrase "evolutionary humanism" in the decades before he coined the new word). A variety of individuals and movements were part of what would come together under the umbrella of transhumanism, often borrowing from 19th century Russian ideas and all drawing on the emergent consensus of scientific research. Science did not vanquish religion, but it helped spawn new religious worldviews.

Among the contributions of geologists and biologists was, crucially, a sense that the world had changed and was continuing to change. These implied, of course, that the future would bring new opportunities. And so a literature arose around what was and what could be. This literature was wide-ranging, extending to scientific research, such as by Ilya Ilyich Mechnikov (aka Élie Metchnikoff), who sought to extend the healthy span of human years to encompass the entirety of a person's life. But it also meant far more speculative guesses. In 1921, George Bernard Shaw fantasized in a five-cycle play about humanity evolving into a state of near-immortality (subject only to accidental death) with the prospect of evolving further into a state of pure thought, freed from mortal bodies altogether.

The influence of this changing conception of humanity was particularly profound on two friends who would alter the face of both science and the way we think of humanity: John Burdon Sanderson (JBS or Jack) Haldane and Julian Huxley. Among other things, Haldane and Huxley were biologists who contributed to the neo-Darwinian synthesis of the early 20th century. The basic nature of that synthesis was the combination of Gregor Mendel's laws of inheritance with evolution by (primarily) natural selection. Understanding how inheritance worked made it possible for evolution to make sense, which it did in dramatic fashion. Within that intellectual environment, Haldane helped develop the field of population genetics and Huxley contributed to anatomy, embryology, evolutionary systematics, and to the idea of biological conservationism. Huxley, the brother of writer Aldous Huxley, wound up as the first Director General for the United Nations Educational, Scientific, and Cultural Organization (UNESCO) while Haldane moved to India and became a great popularizer of science there even as he continued his studies of biology. There's a darker side to Huxley's and Haldane's approach to human genetics, one we will return to in chapter six. But for now, I wish only to explore their contributions to our biological self-understanding.

Haldane and Huxley promoted the future transformation of humanity into something new and greater, though they did so from different vantages. Haldane took an explicitly anti-religious approach while Huxley, who shared Haldane's atheism, took the other path. It doesn't take belief in god to have a religion! The impact of these two would be profound, and their legacies intertwined. Much of 20th century transhumanism followed Haldane with explicit condemnation of religion; but the religious side of transhumanism grew as the century turned over. Today, these two perspectives are so closely connected that pulling them apart has become functionally impossible.

Haldane began his own transhumanist thinking as a schoolboy. Samanth Subramanian notes in his biography of Haldane that, in his youth, the biologist had read Mechnikov. In his notebooks, Haldane speculated that humanity might become immortal in the distant future. By the 1920s, he was ready for the philosophical broadside that would shape transhumanism to come. In a speech to the Heretics Society at the University of Cambridge, published as *Daedalus, Or, Science and the Future* (1924), Haldane argued that science would replace religion and that the future would bring various wonders, including solar and wind energy, neuropharmacology, and, perhaps, genetic engineering (though of course he did not have the phrase "genetic engineering" to describe it). In a world where scientists could invent a new humanity and a new society, he proposed that they also become god-killers. He wrote "there can be no truce between science and religion" and that those "in whom reason has become the greatest and most terrible of the passions" would be "wreckers of outworn empires and civilisations, doubters, disintegrators, deicides." Even a century back, futurists wanted to "move fast and break things."

Haldane may have felt humanity needed science to vanquish religion, but Huxley saw the need for a "new idea-system" or a new "myth of human destiny." In *Religion without Revelation*, first published in 1928, not long after Haldane's publication of *Daedalus*, he argues that "what the world needs is an essentially religious idea-system, unitary instead of dualistically split, and charged with the total dynamic of knowledge old and new, objective and subjective, of experience scientific and spiritual. This is not merely desirable but urgent." Following upon the end of World War I (and revising the book in the wake of World War II), Huxley saw a world upended, in part through crises enabled by technology. At the time, he held out the hope that we could handle our newfound estate by building up a new worldview. He wrote that

> Twentieth-century man, it is clear, needs a new organ for dealing with destiny, a new system of religious beliefs and attitudes adapted to the new situation in which his societies now have to exist....But the need to-day is for a belief-system adapted to cope with his knowledge and his creative possibilities; and this implies the capacity to meet, inspire and guide change.

Whatever dream Huxley might have had for a modern religious system, we have thus far failed to realize it. Whether a transhumanist religion is the right answer, I do not pretend to know; but I suspect that if it is, we need something other than the versions we see coming from Elon Musk, Sam Altman, and Balaji Srinivasan (to name a few people whom we'll discuss in this book).

In his 1957 book, *New Bottles for New Wine*, Huxley makes a special case for the necessity of this new, technologically inspired religion. He argues that we need a religion in order to carry us into the future, an ideology that we can believe in and use to transform ourselves and the world. And it was in *New Bottles for New Wine* that Huxley again used the term transhumanism, this time solidifying its impact.

Inspired by the Jesuit scholar Pierre Teilhard de Chardin, Huxley argued that the universe is "becoming conscious of itself." As the integral part of this – the actualization of the universe's consciousness – humanity has taken up a "cosmic office" and has an "inescapable destiny": humanity is now in control of the future of life on Earth. Thus, the religion that Huxley proposes is not simply one of redemption but one of responsibility. This sense that human beings have work to do is consistent with Huxley's devout commitment to natural and cultural preservation.

Our cosmic office is a critical aspect of 21st century life, even if it gets too often ignored. Thanks to our understanding of geological and biological history, we can now recognize that humanity has a precarious place in the universe. We were not always here, and there is no guarantee that we will always be here. And so, we look to the future and know that for humanity to persist and thrive, we unquestionably must accept – indeed we have no choice – the responsibility that Huxley describes. We have a job to do, one that so-called "longtermists" hope to engage.

Decades before "longtermism" was the rage in Silicon Valley, Robert Ettinger spoke of the Long View and he dismissed, like Huxley, the religion-science conflict espoused by Haldane. Ettinger wasn't out to create a religion; but he felt that the Long View was consonant with traditional religions. In *Man into Superman*, he writes that religious people "will soon see that we are really looking for an ultimate ecumenism, a final rapprochement between science and religion." Most modern longtermists would probably put themselves in the Haldane tradition (if they're aware of the distinctions), but the trajectory remains part of Huxley's.

Longtermism argues that we should make choices based on a calculus that includes all the potential people of the future. If we imagine a spacefaring humanity, there could be trillions of human beings of some distant point in time. Supposedly, then, a decision made that benefits them at the cost of contemporary human beings would be ethically correct. Longtermism resists the short-term model of most decision-making, and that's a good thing. It might represent some move in the direction of Huxley's cosmic duty.

Unfortunately, the longtermist movement, even before it was badly tarnished by the fraudulent philanthropist Sam Bankman-Fried, wasn't likely to get the job done. The chief philosopher of longtermism, William McCaskill, seems to be a thoughtful person: the view that we should contemplate our decisions in terms of future generations is a fundamentally sound one and reflecting on future consequences is important. Too often, however, the movement goes astray, proposing that we basically ignore the present in concern for all the alleged people of the future.

We must think about how to protect people and the planet now as part of our contemplation of the future. If we are to provide shelter for humanity in the future, we need a proper worldview of shelter in the present. We require a worldview of mythical proportions, one that does the real work of ensuring the continuation of civilization even amidst the ongoing potential for cataclysm. To prepare for such a worldview, we must also recognize some handicaps built into the futureproofing agenda.

Shelter for all

Beyond the longtermist community, the tech advocates of future-proofing humanity often have a hard time seeing the present. Future-proofing is, by definition, about whether there will be a human species in the future; but when we stare at glorious mountains on the horizon we sometimes miss the potholes right in the road. Some of our existing troubles are already of dramatic consequence; so ignoring them seems counterproductive. By calling them out, we can appreciate where futureproofing's champions begin to get things right.

The redefinition of death is a clear example of how transhumanists and futureproofers often ignore present troubles in exchange for the hope of future achievements. Cryonics, a movement founded by Robert Ettinger in his first book, *The Prospect of Immortality* (1964), is the practice of freezing heads or entire bodies of the deceased with hope that future technologies will permit their resurrection: heads and bodies floating in liquid nitrogen, awaiting their return to mortal life. Cryonics is international in scope, despite Ettinger himself suggesting that in most countries only the most primitive and, he admits, worthless techniques would be available. While western countries might have liquid nitrogen, Ettinger described early African cryonics as utilizing ice and packed straw. He knew full well that no such thing would prove useful. Ettinger could thus have benefited from the contributions of contemporary 21st century transhumanists who have seen a bit more of the world and have gained a more empathetic approach.

Transhumanists coming from developing nations see advanced technology rather differently from those in the safe confines of mid-20th century America. Osh Agabe, a Nigerian-born neuroscientist and entrepreneur, for example, argues that making capitalism address the long term would mean counteracting unjust financial and political systems now. As for transhumanist technologies, he writes that "the first task for technologies that connect any two brains to each other will be to solve the problem of empathy." Perhaps if Agabe gets his way, the strongest among us will see a future path that includes and even benefits the weakest. We will all benefit from his labors as he pushes neurotech toward applications that "are driven not by profit, but by a spirit of curiosity and the need to know ourselves and to live with others."

Similarly, the African philosopher Fayemi Ademola Kazeem suggests that the value of each individual in society is based on their moral and social contributions to the group. Drawing on Yoruba traditions and African philosophical argument, he argues that personhood – the state of *being* a real person – is "an unending process of becoming more of a humane person, acquiring virtuous attributes." In his understanding of Yoruba thought, a human being can become *more of a person* and does so by becoming more helpful to those around themselves. He suggests that we could use biotechnology to make the species "more 'humanly humane.'" Genetic interventions to enhance our moral dispositions would impact how we reflect on others, and perhaps remove the mental stumbling blocks that allowed Ettinger to cheerfully envision straw-packed Africans and neatly resurrected Americans.

The hope cryonics offers to its practitioners is fundamentally religious: future salvation in which people are not only restored to life but offered a better world in which to live it. The world will have solved the problems of sickness and old age and be brightened by persistent marvels. His own faith confirmed, Ettinger followed *Prospect* with *Man into Superman* (1972), a book that helped underwrite modern day transhumanism with its promise of augmented humanity and immortality attained.

Cryonics is really too individualistic to play a significant role in the futureproofing of humanity, but it has penetrated into the transhumanist space and from there into wider public awareness. According to a survey undertaken by Ariel Zeleznikow-Johnston and his collaborators, around 40% of neuroscientists believe that "successful whole brain emulation could theoretically be created from the structure of a preserved brain." This means that anyone whose head was cryonically frozen could later participate in some form of transhumanist resurrection through computer simulation (the digital emulation of the mind). When the survey caught the attention of news media in December of 2024, the online archive for psychology preprints could not serve all the digital requests coming in for the published paper, causing download failures. There is definitely an audience for technological resurrection, though the inability to offer robust article download capacity might cast a shadow on the technological likelihood of emulating human brains.

Given the apparently widespread interest in personal salvation as part of futureproofing, promoting empathy will be a necessary part of the worldview. Unfortunately, the early advocates of Haldane's atheist transhuman legacy walked right down Ettinger's path, though often unwittingly. I think it is safe to say that when FM-2030 (né Fereidoun Esfandiary) argued for "cosmic rights" he envisioned all humanity receiving them. By all accounts, he seems to have been a fairly decent person. But only someone who has never needed to worry about how to keep a roof over his head can say "we are no longer content with building shelters for the homeless...We are on the way to eliminating the very concept of fixed shelters, homes, towns" or "we can never again be content with civil rights...We now want cosmic rights."

Let's get shelter and civil rights for everyone first. Then maybe cosmic rights, whatever those may be.

The privileged, if accidental, disregard for the real plight of human beings, sprinkled with a liberal dose of Ayn Rand's deliberate disregard for them, led FM's followers down a problematic road of libertarian self-interest. Late 20th century transhumanism, as it emerged in the Extropian movement led by Max More (né Max O'Connor), Natasha Vita-More (née Nancie Clark), T.O. Morrow (né Tom Bell), and others pushed such a hyper-individualistic perspective that, if implemented, it would likely have spelled the doom of transhumanism and humanity both! The democratic transhumanism led by practicing Buddhist James Hughes re-injected compassion into the conversation and provided a much-needed intervention. But in the early days, indifferent libertarianism and a complete rejection of shared public policy reigned. Illustrating this, Romana Machado often attended Extropian events cosplaying as "the State" ... a dominatrix dragging a crawling "taxpayer" along at the end of a leash. Over time, most Extropians backed off their callously anti-government agenda and considered instead the possibility that all people should be supported in their effort to grow and "progress."

This is quite in keeping with Huxley's desire to see humanity reach "fulfillment." He saw fulfillment as an individual and collective project, one where we could fully pursue our own ends only insofar as those ends were connected to the benefit of the species and of the world. There is no merit to the idea that humanity could overcome existential risk if we reject collective responsibility. "The human species can," he writes, "transcend itself—not just sporadically, an individual here in one way, an individual there in another way, but in its entirety, as humanity."

Religiously posthuman

Another way Huxley has outlasted Haldane is in his explicit (if atheistic) religiosity. While the early 21st century was rife with transhumanist attacks on "dogmatic faith," as in the Principles of Extropy, it took only a decade or two for influential sociologist William Bainbridge, retired physicist Giulio Prisco, new age spirituality communities like Terasem, and others to rejuvenate the religious affect of transhumanism. The influence of Lincoln Cannon, who launched the Mormon Transhumanist Association, was particularly key because he found common ground between a traditional religion and transhumanist thinking. Thanks to that work, Christopher Benek launched a Christian Transhumanist Association and other groups with religious connotations, like India Awakens and Theta Noir, moved onto the edges of public awareness and conversation. All of these individuals and movements recognized that there is something inherently religious about promising cosmic purpose and individual salvation. Such religious goals don't make transhumanism wrong or foolish or bad. These goals, whether attainable or not, are perfectly natural aims that people have transferred from "old time religion" to science and technology.

Turn of the 20th century Russia was a watershed moment in the convergence of religion and technology, and these contributions show new life in the religious transhumanism of the 21st century. The provocateur of the new worldview was Nikolai Fedorov (1829-1903), a librarian in Moscow. He and his Russian Cosmist followers made the resurrection of the dead and delivery of immortality the core principles of what he called the Common Task. If death and taxes are the only inevitables, then surely ending death is a shared project. At least, that is how Fedorov thought about it. He linked this goal with the inauguration of universal kinship and the end of war. For Fedorov, that effort was explicitly Christian. He felt that Jesus intended for humanity to establish universal brotherhood (women don't figure into his thinking very often), personal immortality, and resurrection of "the fathers" tracing all the way back to Adam. When we discuss spaceflight in chapter five, we'll return to Fedorov, for he believed the Common Task could be completed only by traveling among the stars.

Early in the 21st century, largely thanks to the work of George Young, western scholars and transhumanists simultaneously rediscovered the Russians of the early 20th. Following on dramatic growth in transhumanism and its cultural impact, Young – a scholar of Russian history – resurrected some of his old studies to produce the first English-language study of Cosmism, a movement that affected everything from literature to rocket science in the Soviet Union. Fedorov's ideas even influenced Bolshevism in the early U.S.S.R., with dreams of immortality coursing through the public speeches of leading figures like Leon Trotsky and his colleagues. Though forgotten until recently, the Cosmists also directly impacted Euro-American transhumanism, and we will see some of these connections in the coming chapters.

A growing body of literature in English has exposed Fedorov's ideas to non-Russian audiences. This is the fulfillment of Dmitry Slapentokh's 1996 prediction that Fedorov's beliefs "might reemerge in some other culture, for great ideas recognize no boundaries in time or space." Today, Fedorov's ideas indeed motivate western transhumanists, and the resurgence of Cosmism has prompted 21st century Russians to take up the mantle of technological salvation. Fedorovians generally remain tied to Russian Orthodox Christianity, but other transhumanists in Russia reject those connections and seek an independent movement along the lines created by FM-2030, Max More, and others in the west. Meanwhile, the Cosmist vision directly influenced western transhumanist visions, including AI researcher Ben Goetzel's *Cosmist Manifesto*.

People like Fedorov and Teilhard de Chardin confidently rested in their Christian faith that history moves inexorably toward salvation; but few in the 21st century can be as sanguine. Are we not faced by a mounting list of existential risks? We know the solar system is filled with asteroids (some potentially on a trajectory to collide with Earth) and there are supervolcanoes whose eruptions are geologically impossible to predict. Scientists warn us that overfishing could drive most or all of the ocean's fish species extinct by later this century and that the bee species who undergird most of our agriculture are at risk of extinction. Political partisanship or climate-fueled migration could lead to ever wider and more dangerous forms of war. It is with clear justification that a pessimist might look upon the world and forsake all hope.

And yet futurists claim that our salvation is assured, or nearly so. FM-2030 was a bedrock of such optimism: he believed that the future, even the present, brings wonders unheralded by the past. He argued that humanity had no reason to be pessimistic anymore because we have the power to solve problems once intractable. His cryopreserved head floats in liquid nitrogen as a testament to his faith. But religions work better when they come with guarantees. What is there to assuage our fears when we face so many potential cataclysms?

Teilhard de Chardin thought evolution aligns with Christian beliefs about the future of the universe, that history moves straight toward its own salvation. But as we saw early in this chapter, the secular salvation of science rejects the safe promise of divine intervention. Instead, faith gets lodged in the scientific claims themselves. Take, for example, the eminent roboticist Hans Moravec. In *Robot: Mere Machine to Transcendent Mind*, Moravec states that evolution is "weeding out ineffective ways of thought" and that the consequent arrival of super-intelligent robots is "inevitable."

When we think about the direction of the universe, we must ask what motivates movement in that direction. For Christians, there is a divine mandate in the linear movement of history. They inherited this, of course, from Judaism, in which the world had a moment when it came into existence and – according to many – would have a final moment of judgment. Not all Jews anticipated or described an inevitable cosmic conclusion; but enough did so around 2000 years ago that Jesus of Nazareth made oblique references to the possibility. One of his later followers, who would change his name from Saul to Paul, lived in firm conviction that Jesus would return from death and end the world in Paul's own lifetime: "We will not all die," Paul writes in a letter to the Christians of Corinth, "but we will all be changed." And to the Christians of Thessalonica he writes that when Jesus returns to Earth "we who are still alive and are left" will join the risen dead to "meet the Lord in the air."

The futureproofing crowd draws on the same religious faith in imminent salvation. It's not far from Paul to Robert Ettinger, founder of cryonics, who alleges in *The Prospect of Immortality* (1964) that "most of us now breathing have a good chance of physical life after death – a sober, scientific probability of revival and rejuvenation of our frozen bodies." Nor is it much distance to Kurzweil writing in *The Singularity is Near* that "most of the readers of this book are likely to be around to experience the Singularity" when machine intelligence makes miracles possible.

Generally speaking, the futureproofing movement rejects an eternal god directing the course of history toward its culmination. There are transhumanists who ascribe also to a traditional religion (e.g., the Mormon Transhumanist Association and the Christian Transhumanist Association), but it is far, far more common for transhumanists to reject traditional religions and their gods. And yet, while they have excised gods from their worldview, they remain committed to many of the cultural components associated with those gods. A linear path of cosmic destiny is one of those, and it now rests on the success of modern science.

Evolution is a common resource for those who seek transcendent guarantees, but this is a double-edged sword. Such is the route taken by Moravec, who earned his PhD at Stanford, spent his academic career at Carnegie Mellon University, and was one of the seminal figures in mobile robotics. He believes that "intelligent machines, however benevolent, threaten our existence because they are alternative inhabitants of our ecological niche." It is unclear what niche he means, given that human beings and machines have differing needs for their habitations, energy supplies, reproduction, etc. In fact, human beings and machines have such wildly different ecological niches that if *that's* the reason we are threatened by machines (and we will return to his answer to this in chapter three), then we should feel quite safe! Nevertheless, by putting us into competition, he can justify the evolutionary transition of human beings *into* machines. That is, evolution seems to guarantee a future world in which biology ("mere jelly," in Moravec's words) will cede primacy of cosmic place to machines. This process – because evolution is said to work on cultures and concepts as well as genes – now appears inevitable.

In reality, of course, we just use the *word* evolution to refer to cultural change, technological progress, etc. This is actually a disservice to Darwin's own construction of biological evolution, but it has become commonplace. The use of evolution for these purposes promotes a new sense of cosmic direction on human culture and the objects we build. Words shape the world, and in this case using a specific word to talk about change allows that word to take on added strength in public conversations about our path into the future.

Evolution may not mean quite what Moravec or others want it to mean, but transhumanists have alternatives. Most famous of these is Ray Kurzweil's law of accelerating returns. According to Kurzweil, the rate of progress in between "salient events" always accelerates exponentially in ordered systems because order itself does so. An ordered system is, of course, any that is not random in its construction. For example, the Earth is an ordered system even though the wider context of the universe it is not. The influx of sunlight is the primary energy source for our biosphere, preventing the kind of entropic breakdown predicted by thermodynamics. Instead of increasing chaos, we see increasing order (only locally...the universe is still collectively becoming more chaotic and the energy we derive from tectonic forces and solar radiation is temporary in cosmic timelines). Kurzweil looks at the 4.5 billion year history of Earth and rightly sees increasingly complex and orderly systems and structures. He then proposes without evidence (because there can be none) that such ongoing progress is a fundamental property of the universe: ordered systems will always exponentially increase in their amount of order forever.

The exponential increases Kurzweil tries to document lead to doubling rates of progress. That is, for every passage of a specific period of time, the amount of order should double. As a clear example, Kurzweil relies on Moore's Law. Named for Gordon Moore, who described it in the 1960s, Moore's Law states that you can double the number of transistors on an integrated circuit approximately every 18-24 months (until such point as quantum interference makes smaller circuits impossible). Doubling those transistors doubles the speed of the integrated circuit and thus of the computer. So what Moore described was the fact that every year or two a brand-new computer would be twice as fast as the previous model. Those of us living through the past 30-40 years witnessed something to this effect. Kurzweil expands Moore's logic to the entire universe. Just as the speed of computers could double, he says that it is fundamental to order (as long as you have an influx of energy) to create more order.

The impact of exponential growth is immense. If I double a very small number, say .00001, I get another very small number (.00002). But if I keep at it, I will eventually reach somewhat more noticeable amounts; let's say ¼. If I double ¼ all I get is ½, which is only somewhat larger. But I double that and get 1, then 2, then 4, then 8. Eventually, in the same time period that I turned ¼ into ½, I manage to turn 256 into 512 and subsequently from there to 1024. You can see that eventually this means staggering changes, especially if it is neverending. As the son of an economist and an accountant, however, I have resource management in my blood. I know that no exponential curves last forever. Advocates of what is called the Singularity disagree. They believe that the curves have no meaningful upper bounds. Kurzweil seemed to hold this position until *The Singularity is Nearer*, in which he revises his approach to exponential curves (I will explore this change of perspective in chapter three). Whether it is bounded or not is less important, however, than where the upper bound lies. If we will reach it soon, then there will be no single moment of cataclysmic change. But if the exponential progress of technology will continue for a while, then there will be a singular moment in time where the entire universe shifts from one reckoning of time, intelligence, and life to another.

The Singularity, then, is the point at which the doubling of computer capabilities has massive consequences for computer "intelligence." If we assume computer speed is somehow the same as intelligence (which is doubtful) then we double our way to the point that a computer is similar to a human and then swiftly double our way far, far beyond the human. The Singularity is that moment where progress from a single doubling is simply impossible to comprehend from our current vantage. I can predict what life will look like in two years, and I'll be reasonably accurate. But if the world really is subject to exponential logic then there will be a time when we cannot even predict what the world will look like two years into the future. If that process does continue to the point of tremendous technological progress, then humanity will witness the emergence of new, super-human forms of life. With appropriate application of futureproofing technologies, humanity will not be overcome by superhuman computers, instead joining them as equals and partners.

And so, without use for any gods at all, futureproofers say that we can look toward a magical era to come. If we choose wrongly, we will simply go extinct. If we choose wisely, we will live past the existential risk of evolutionary out-competition and join this remarkable future.

The religious remnant

From the late 19th to the early 21st century we have seen specula-tion about the future: ghostly dreams of what could be. Those who aspired to immortality missed the mark, though cryonic suspension may provide them with another run at it. The emergence of transhu-manist speculation preceded the marvelous scientific breakthroughs in spaceflight, genetic engineering, and artificial intelligence. The rise of these technologies cannot be fully explained without recourse to the futurist promises of human transcendence, and those promises themselves were reinforced and, occasionally having gone dormant, reinvigorated by technological achievement.

From Fedorov's Common Task to FM-2030's philosophy of opti-mism, a consistent stream of technological utopianism characterized the 20th century. This is not to say that everybody was sanguine with either our technologies or our future. In fact, the academic study of religion and science largely emerged out of fear that nuclear weapons would be the end of us. With many technological advances, the fear of upending nature or upsetting the gods has risen to critique them. But for all the loud concerns, many of which occupy chapter six, few people stand resolute against the siren call of technological progress. Collectively, we *do* believe that we can enhance our lives with technology. And so, when transhumanists cheer us onwards they find a ready audience. Transhumanism is the 21st century ethos of technology.

To gaze upon entrepreneurial technology is to witness dreams of transcendence. Profit, yes. But also transcendence. For while someone stands to make money from longevity treatments or advanced AI, the justification for those pursuits is fundamentally religious: they are the route to human salvation.

CHAPTER TWO
ZOMBIE RE(GENE)RATION

Whether to overcome the hardships of fossil fuel burning, the inevitability of earthly catastrophe, or just out of a refusal to accept aging and death, futurists dream of bioengineering humanity. Such ideas emerge, run their course, pass away, and rise again. These opportunities to reshape our species and resist biological death emerge from our scientific understanding of Darwinian evolution, Mendel's laws of inheritance, and the chemical pathways of DNA: when scientists identified the DNA coding for an organism's traits, they gained an inkling as to how they might change that organism permanently. Scientific understanding and technological developments in genetic engineering thus produced a slow burn of imaginative promises in biology, redemptive promises that range from a restored Pleistocene-era grassland to biotech-enhanced super-humanity.

History buffs know there is nothing particularly new about the search for biological perfection or the integration of science and technology toward it. European Christians spent centuries thinking that science might restore them to a state of paradise. In *The Religion of Technology*, David Noble traces this effort, showing that for many Christians, the increasing power and knowledge enabled by science would bring humanity back to its original state (i.e., before eating the fruit of knowledge of good and evil). As science progressed, Christians assumed that Adam must have possessed all the newly acquired knowledge prior to eating the fruit of knowledge. These men never seemed to suppose that *Eve* possessed all scientific knowledge, instead confidently asserting that she was just the cause of Adam's fall from a state of intellectual as well as spiritual grace.

But whether we include Eve or not, the Christians of the second millennium argued that Adam knew everything that could only lately be learned by science. They supposed that even the powers afforded by technology were possessed innately by Adam. Innovations like the telescope and the microscope returned our sight to the level of Adam's before the Fall. So the pursuit of scientific knowledge and technological empowerment were nothing more than a return to the glorious days of wandering the Garden of Eden.

The explicit Christianity of all this faded, especially in the late 19th and into the 20th century. Where once knowledge of the natural world might be a divine plan to restore humanity to oneness with god, it came to mean no more – but also no less – than the opportunity to overcome human limits without divine help. Science became, as philosopher Mary Midgley writes, salvation.

The secular era that produces these new technological religions is, itself, the curious outcome of early modern comparative religion. For it was comparative religion, itself an exercise of colonial domination, that helped destabilize the very religion (Christianity) which it served. In his book *Savage Systems*, David Chidester shows in concrete detail how Europeans derided foreign customs as savage, animal, or inhuman, and then used this as their justification to conquer and enslave. In particular, by declaring that Africans lacked religion, Europeans argued that there was something inherently less than human about them. In *The Empire of Religion*, he explores how the study of various religions emerged as Europeans found new trading partners and, ultimately, decided to conquer as much of the world as they could. Initially, comparative religions aimed to bring the entire world into a single worldview; but this attempt failed.

Instead, the comparative enterprise exposed Europeans and (those they sought to conquer) to a world of rich religious diversity. Although initially allowing Europeans to colonize the minds as well as the lands of the world, it ultimately opened people's eyes to the possibility that many views could be correct or, in fact, that they might all be wrong. If the religions are, at least in their details, incorrect, then they cannot lay sovereign claim to all history. They lose the power to determine how we perceive the origins of the Earth, the emergence of humanity, and the role of humanity in the cosmos.

The study of religion thereby set a stage for the eventual rise of today's technological religions of existential risk. In a world where no traditional religion can claim authority over biology, it came about that biology claims authority over religion. This happens in neuroscientific efforts to explain religion, such as in dreams of a so-called "god gene" that predisposes people to religious experience. It also happens in cognitive science explanations, such as the agent-detection model, which proposes that human beings evolved to detect agency in the world (for lots of good reasons having to do with being both hunter and hunted) and thus project spiritual agency on accidents of nature and chance. Throughout this book, we will see how other scientific efforts generate religious fervor, such as how biology underwrites a religion of genetic superhumanity.

Christians have dreamed of immortality for two thousand years, and their dreams of transcendence are now firmly ensconced in commercially marketed biology. What I mean is that there are now enormous financial resources poured into start-up longevity businesses designed to unlink human life from age and death. The people committed to that enterprise appear throughout the future-proofing community. For example, Balaji Srinivasan, to whom we will return in chapter four, is a financial backer for Biograph, the "world's most advanced preventative health clinic." Ray Kurzweil, a key figure in chapter three, was ahead of the curve: in addition to his books on AI and the Singularity, he is also the co-author of *Fantastic Voyage: Live Long Enough to Live Forever*, which he published in 2004. He accompanied that book with an entrepreneurial effort to sell vita-mins and supplements to his faithful.

Taking up this dream, those who would turn immortal in the 21st century find much to love about advancements in biotechnology. Most famously, tech entrepreneur Bryan Johnson builds a transhu-manist practice out of blood transfusions, dramatic consumption of supplements, facial fat implants (to unsuccessfully compensate for his near-starvation diet), and even electrocuting his penis. In the early 21st century, longevity research acquired just the barest hint of respectability after decades on the counterculture fringe. By 2025, doctors could lay claim to prestige by associating with Silicon Valley capitalists desperate to extend their economic and social privilege into the indefinite future. Entrepreneurs whose own work might be far afield from biotech, such as OpenAI's Sam Altman, invest millions of dollars in longevity companies. Simultaneously, their desire for immortality makes solving the existential risk crisis a high priority for them.

Ultimately, the movement to reach genetic transcendence is a secular mirror of traditional religious concerns. It is a vision of human beings resistant to both individual and species death. The reader might already expect that all these potential shifts in human biology, psychology, and culture were present as fringe possibilities until the watershed moment when existential risk became the perspective *du jour*. Like zombies, they rose up from the grave at this new opportunity.

Zombies, of course, are notoriously hard to kill: they just keep coming back around. They are metaphors for consumer capitalism, as in Romero's *Night of the Living Dead*, and, as I've argued along with my former students Nat Recine and Samantha Fox, zombies also domesticate our psychological terrors. Zombie DNA is actually a real thing also: it's leftover DNA from ancient viruses that lies dormant within a person or species, and it can resurrect and reassert itself in an organism. But in this chapter, what ties biotechnology to zombies is their pursuit of un-death (not so much the undeath of horror movies, fortunately) but also the fact that the biotechnological dream of immortality got buried, but never exorcised, from the hopes of biological sciences. By the late 20th century, the dream started clawing its way back into the light. Early in the 21st century, the same advocates of perfect cities and AI futures also throw their weight and prestige behind biotech companies devoted to genetic enhancement and eternal youth. And so futureproofing humanity begins with geneticproofing.

Life, unlimited

The end of aging is an essential part of the 21st century reconstitution of our biology. Obviously, there's a conflict between living forever and some of the existential risks that concern the futureproofing community. In particular, if we do not die of old age, we risk exacerbating our impact on the climate, increasing pollution to unsustainable levels, and wreaking havoc on plant and animal species. So, it seems that eternal youth is counter-indicated if we hope to sustain life on Earth. Such pressures do not, however, undermine the goal of immortality; instead, they invigorate the rest of the futureproofing agenda. The occupations of later chapters, especially chapter five on spaceflight, respond to the full panoply of existential risks but also to the problems suggested by the pursuit of eternal life.

Human beings have tinkered with immortality for centuries, but recent efforts are likely to be more substantive than the alchemical pursuits of the past. The religious search for immortality in this life (as opposed to the typical pursuit of immortality in a next life) includes everything from Daoism in ancient China to chemical laboratories in early modern Europe. Certainly, alchemy has other goals as well; but the pursuit of immortality is at the root of alchemists' experimentation.

Paracelsus, Nicolas Flamel, and other Europeans are the names most closely tied to alchemy in the western context. Flamel (c. 1330-1418 CE) wasn't likely even an alchemist, or much of one at any rate; but he earned a reputation as such centuries after his death and is now globally famous thanks to a certain young adult fantasy series.

Paracelsus (1493-1541 CE), on the other hand, certainly sought alchemical immortality, which explains his probable death from mercury poisoning (he consumed the toxic substance "medicinally"). Where Flamel was only posthumously connected to wonders like the philosopher's stone, which would permit transmutation of lead into gold and the creation of an elixir of life, these were the province of Paracelsus and a part of his reputation as a medical doctor.

Alchemists promised more than eternal life; they also sought to create life, imitating the divine. The Islamic alchemist Jābir (c. 721-815 CE) made no attempt to become immortal (which would have flown in the face of Muslim theology) but offers among the earliest known descriptions of an homunculus, a living creature crafted through alchemical labor. A later Islamic alchemist authored *The Book of the Cow*, attributing it to Jābir, and includes instructions for creating an homunculus with magical powers ranging from walking on water to affecting the phases of the moon. Paracelsus, no doubt informed by such speculation, also claimed that an homunculus gave super-human powers to its creator. The creation of life, the acquisition of superpowers, and the search for immortality: alchemy is an important part of the intellectual ancestry of transhumanism.

It's no accident that alchemical immortality is having a moment in worldwide public life. Scholar Zhange Ni states that in the 21st century, Chinese readers increasingly turn to "immortality cultivation fiction...a uniquely Chinese type of fantasy that builds imaginary worlds around the magical practice of Daoist alchemy." That genre of literature builds on centuries of Daoist thought in China, but also takes advantage of 20th century transhumanism, especially the work of prolific innovator Ray Kurzweil, who was particularly well-regarded in the 1990s and early 2000s. Ni notes that Kurzweil's *The Age of Spiritual Machines*, a key book in the spread of transhumanist thought (to which we will return in the next chapter) was translated into Chinese early in the new millennium, from which point it powerfully influenced literary culture. The combination of alchemy and transhumanist thought in Chinese literature proposes a religious and magical approach that overcomes some of the libertarian and culturally exclusive versions of technological salvation we saw in the last chapter. In the present, then, perhaps there are lessons emergent in our traditional visions of immortality.

Before transhumanists reckoned they could unite science, technology, and immortality, however, a few scientists found their way to a middle ground. In spite of real and potential criticism, visionaries and scientists occasionally promised semi-realistic wonders for our biotech future. Since the 19th century Russian scientist, Ilya Mechnikov, there have been occasional scientists committed to finding a cure for aging, to scientifically reverse and perhaps permanently halt the process. Mechnikov studied naked mole rats, which have a unique aging process – they remain in full health throughout their lifespans and then die. He felt that human beings could be altered in such a way as to thrive until their 120th year before dying. His work influenced a young Jack Haldane, who saw it as inspirational toward a bigger vision: one in which human beings simply lived indefinite lifespans. Mechnikov was also surely known to the founder of Russian Cosmism, Nikolai Fedorov, who believed that it was the obligation of humanity to overcome death and resurrect all preceding generations – "the fathers" – going all the way back to the biblical Adam.

Mechnikov saw no reason why human beings couldn't live – like naked mole rats – an extended healthy life before dying and without ever suffering the effects of aging. That perspective failed to catch fire until much later, with a century of scientists largely avoiding talk of ageless humanity and life extension. Instead, medical science favored disease prevention. Russian Cosmists built upon Mechnikov and JBS Haldane cited him favorably; and yet western science largely looked the other way when anti-aging gurus argued that vitamins and yoga could extend human lifespans.

Today, the pursuit of human longevity still draws on those turn-of-the-20th century Russians – some of whose rather odd ideas have become current once again. George Young notes that Alexander Bogdanov "died as a result of his own medical experiment, probably from injecting himself with blood of an incompatible type." It was

Bogdanov who first thought that transfusing blood from a younger person would reduce the age of the older. When PayPal entrepreneur Peter Thiel resurrected this goal, he was immediately, and hilariously, lampooned as a would-be vampire drinking the blood of the youth. Like Bogdanov, however, Thiel (and Bryan Johnson mentioned above) supports blood transfusion, not vampirism. Whether the current longevity experiments come with risks remains to be seen, though at least blood and plasma transfusions are safe now. It is unlikely that the anti-aging crowd will suffer Bogdanov's fate.

Of course, in the case of some biotechnologies, their mere premise seemed absurd to scientists. The inevitability of death surely dissuaded many scientists from engaging such challenging ground as the halt and reversal of aging. Not only did it seem impossible but, if anything, it seemed that acquiring funding would be even more so.

Nothing is forever, however, and the allure of entrepreneurial start-up cash among scientists, rapidly improving technological knowledge, and growing familiarity among the general public made extravagant scientific promises easier to make and to believe. The 1990s and early 2000s produced a biotech bubble in which eager investors, many newly rich from the previous tech bubble and eager to stay that way in perpetuity, leaped to support anti-aging research. Biologists like Cynthia Kenyon, who quintupled the lifespan of *c. elegans* worms, powered the entrepreneurial enthusiasm. The fringe movement of taking endless vitamin supplements like co-enzyme Q10 ("ubiquinone") and fending off free radicals in our cells suddenly looked to have a few knights in shining armor. For example, Roy Walford followed the early 20th c. work of Clive McCay to get mice living far longer than average by restricting their calorie intake. And so what was, for all intents and purposes, a scientific graveyard (or a commercial shell game), suddenly became a potential biotech wonder. Perhaps with the right approach to human genetics, we could end the "tyranny of age." By the 2020s, companies and fundraising emerged for lifespan extension in people and even our pets. The XPrize Healthspan program, for example, created a $101 million prize purse to *reduce* a person's age (as measured across musculoskeletal, cognitive, and immune functions).

If our species can exist in perpetuity, so might the individual. The existential risk crisis that measures whether human beings will be on Earth centuries or millennia into the future hits closest to home for those who believe they have a reasonable chance of living indefinitely. What was once the province of fringe Russian science eventually hit the Silicon Valley mainstream when biotech research offered the first meaningful interventions in nonhuman species. If worms and mice can live longer, why not humanity?

While the cocktail of anti-aging technologies goes from siphoning the blood plasma of young people to cryoshocking the body with ice baths, the most obvious place to intervene is our genetics. There are clear impacts of aging on our chromosomes (e.g., the shortening of telomeres, which are protective DNA sequences on the end of the chromosome) and the existence of diseases that only appear with age show the connection between lifespan and our genetic inheritance. And so the pursuit of longevity escape velocity – when each year medical science restores more youth than the twelve months we spent – draws directly on genetics research. And, while we're mucking around with the DNA to make ourselves live longer...well, ask the futureproofers, why stop there?

Evolution unbound

Futurists testify that ours is the time for humanity to surpass the natural constraints that governed all prior eras of cosmic history. Geneticproofers and other transhumanists believe that we can replace the blind guidance of evolutionary progress with deliberate human choice. There is no need to accept the accidents of fate once humanity fully understands the genetic code and knows how to change what we don't like.

In the last chapter, I raised Julian Huxley's belief that humanity had taken on a cosmic office. It is no surprise that this idea would emerge from a biologist. Huxley wasn't just a spectator to the theory of evolution, he was a key contributor. His contributions to science and culture range outside the neo-Darwinian synthesis (the combination of evolution by natural selection with Mendel's laws of inheritance and, ultimately, the chemical understanding of DNA). But a crucial part of his biological studies contributed directly to that synthesis. He was deeply immersed in the theory of evolution, and thus the idea that human beings had evolved from some prior state. This is a vital perspective from which to launch his ideas about what human beings could become.

After all, if we had emerged from a prior species then it stands to reason we would transform into something else in the future. There was simply no reason, looking at the history of earthly evolution, to believe that human beings had reached some sort of pinnacle of evolutionary fitness. Conditions would change, and we would change with them. But Huxley was absolutely clear about one difference: where once our evolution happened by accident, we were now in a position to make choices about it.

Huxley's belief that we could direct evolution was, unfortunately, his entry point to eugenics. Early on, and subsequently for the rest of his career, Huxley shared the eugenics perspective that some people were more fit to breed than others. While he was no Nazi, he did think that "good breeding" could be encouraged, and that the elite among humanity (however he defined that) had better cause for producing children than others. Fortunately for his legacy, Huxley at least rejected race-based eugenics.

Alongside, and in tension with his eugenics stance, Huxley promoted a vision of personal and collective "fulfillment." He supposed that human beings could pursue their own individual fulfillment even as we pursued it collectively. That perspective was crucial to his feelings about the destiny, the inescapable destiny, of humankind. For if human beings had gained control over our environment and even ourselves, then we stood at the cusp of newfound human self-direction. We were not subject to the whims of nature but could instead turn nature toward our own ends.

It is no wonder that Huxley's vision of evolution percolated within our collective vision of the future and, especially, of our genetic future. While many people remained, and remain, content to allow their gods sole responsibility over the human future, others saw an opportunity to bring about new human – and posthuman – forms of life.

Many scientists were among the naysayers. They feared over-promising and underdelivering. They feared that the challenges were too great. In his wonderful investigation into the biotech bubble of the early 2000s, Brian Alexander describes scientists keeping wary watch on the enthusiastic promises of transhumanists. Outside of science, public cheerleaders charted new futures for human beings, a trend that began by the 1970s. But progress in actual biotech interventions had been painstakingly slow, rich with a variety of risks, and a field of political landmines. Too many of what Alexander calls the "bioluddites" were ready to pounce on scientists for "playing God" and for making dubious claims about the biomedical interventions they proposed. Government officials lambasted them, accusing them of violating the sanctity of life. This was especially pronounced in arguments over embryonic stem cell research and cloning.

The bioluddite arguments were often absurd, such as when Cliff Stearns decided that human DNA has "tentacles" (he meant telomeres); but they resonated with a public that feared scientific displacement of its cherished beliefs. That those beliefs were historically recent mattered little. According to scholar Randall Balmer, it wasn't until the late 1970s that American Protestants decided to oppose abortion, not because they particularly cared about abortion itself but because they could use it as a wedge issue in American electoral politics. *Roe v. Wade* had received widespread evangelical Christian support; but several years later when the US government rejected their ongoing preference for racial segregation at Christian colleges, abortion suddenly became useful to push Jimmy Carter out of the White House. This was a direct response to the IRS removing tax-exempt status from Christian schools like Bob Jones University due to their racist policies. In any case, by the 1990s, abortion politics carried much weight in the United States, and those politics were woven into stem cell research and cast a pall across the entire biomedical enter-prise. Any scientist promising that humanity should alter its genetic inheritance or eliminate limits to our lifespan exposed themself rather rapidly to torches and pitchforks.

We should not sell short the likelihood of the public acclimating to new sciences and technologies. In the 1970s, bioethicists like Leon Kass predicted in vitro fertilization (IVF) would inevitably lead to birth defects and stillborn babies. He declared that anything people found "repugnant," such as producing embryos in a petri dish and injecting them into a woman's uterus, must surely be bad in principle. When generations of children were born without any of the terrible outcomes he predicted, not only did ordinary people become comfortable with "test tube babies," but so did their original detrac-tors. For all his belief that a public distaste (which was more likely just cover for his own) should permanently discourage technological innovation, Leon Kass now accepts IVF in family planning.

This is not to say, however, that everyone comfortable with IVF also supports genetically modifying the embryos formed by it. But there are strong cultural and technological trends, supported by the transhumanist religion baked into society by Huxley and later thinkers, which have humanity in a conversation about how we ought to genetically enhance ourselves. How must we act on our inescapable destiny? Are we obliged to actively rewrite the human genetic code?

Prospective parents already select IVF embryos for characteristics like sex and disease heritability. The latter of these seems like an obvious use of the technology. If children are at risk of inheriting a terrible disease like Tay-Sachs, which is universally fatal by the age of three, then it stands to reason that the parents should select embryos that do not have the disease. Not all diseases are necessarily fatal or debilitating to this degree, however, and so ethical debates emerge over selecting for or against other conditions, such as Down's Syndrome or hearing loss. Furthermore, while many prospective parents have clear preferences about the biological sex of their chil-dren, many outsiders feel that such choices go too far. Compounding this, we can select embryos for even less relevant traits, such as hair or eye color, and it stands to reason that with time we will get better at selecting for other traits. The day may not be too far off when we can select for athletic or musical performance, or for characteristics of physical beauty. There are significant social challenges in terms of whether such decision-making should be available to parents.

But selecting among embryos is not the ultimate goal of transhumanist engineering. If we can identify what Gregory Stock labels "naturally existing constellations of genes," we could actually insert those groups of genes into the embryos, permanently changing a person's genetic inheritance. We would know that such naturally occurring gene sets are perfectly safe because they exist already in otherwise perfectly normal and healthy musical prodigies or athletic superstars. We might also identify ways of enhancing human capacity beyond our present capability. Perhaps we can find genes that will make us faster, smarter, stronger, or funnier than any living human being. Perhaps using a technology like CRISPR – the present best technology for inserting new genes into an individual – we could permanently impact the genetic inheritance of humanity. By modifying the gametes (egg and sperm cells), any changes we make would be inherited by future generations.

For transhumanist advocates, genetic engineering does no more than speed up a process already occurring through evolution. They have been awaiting this opportunity for decades. As might be expected, the promise of genetic changes to humanity was already implied by Haldane, ever ahead of his time. Even prior to the discovery of DNA's chemical composition, the production of proteins through RNA, and the determination of human traits through protein expression (everything from eye color to temper), Haldane foresaw the possibility that we would master the tools to alter and enhance humanity. He might also be the first in a surprisingly long line of futurists to suggest we would be able to do things like give tails to our children. By the time all those other scientific pieces came into focus, the futurist religion of human destiny began to revel in science fiction promises for the future.

The power of science fiction in all this should not be ignored and probably cannot be overstated. Brian Alexander, author of *Rapture: A Raucous Tour of Cloning, Transhumanism, and the New Era of Immortality*, attended an early 21st c. anti-aging conference and saw this power firsthand. When a speaker asked the audience who had also read a particular story by Robert Heinlein, nearly every hand went in the air. "Ah, the Brotherhood," the speaker confirmed. Heinlein is particularly famous for his Lazarus Long novels, which are riddled with strange sexual proclivities and scientific gobbledygook ... and tell the tale of a man who simply could not die of old age. Early in the books it seemed like he simply hadn't done so; by the final novel Heinlein implied that Long would see the entire future of the universe. From entertaining story to aspirational myth.

It was a science fiction story that also prompted Robert Ettinger to invent using cryonics to overcome death. One of the giants of 20th century futurism – admittedly with a modest audience in his day – Ettinger composed *The Prospect of Immortality* (1964) and *Man into Superman* (1972). In the former, he argues that we could freeze the recently deceased and subsequently resurrect them when we find solutions to their causes of death. Rather than have people die of old age, injury, cancer, or other disease, Ettinger supposed that we might simply keep them sufficiently cold and then reverse the damage as part of a restoration to life. While not practical in his lifetime (and perhaps never to be), Ettinger's belief in cryonics spawned a small industry, with companies like Alcor and CryoRus offering services to human beings who financially invest in the hope of resurrection. In some cases, this means freezing an entire body; but in most, the head is severed and frozen, with the presumption that a new body can be cloned, the head can be appended to a robotic body, or the mind can be extracted and uploaded into a machine body.

Beyond life extension and cryonics, Ettinger supposed that the entire genetic inheritance of humanity was open to debate. His visionary work preceded the first actual recombinant DNA technologies in the early 1970s, but knowing how DNA was integral to human gene expression helped him conjecture that human beings could change it. His futuristic vision of *Man into Superman* extended across many domains, from anti-aging to body morphologies to sex/gender dynamics. In many ways, Ettinger was literally decades ahead of his time, prefiguring 21st century understandings of human identity, trans rights, and sexual mores.

Ettinger's contemporary, FM-2030, would not inspire anything directly commercial like Ettinger's cryonics industry; but he became the driving force for the transhumanist project. Once little-known outside of futurist and tech progressive communities, it is likely that FM's name will persist long into the future if even half the transhumanist agenda finds its way to reality. A wide-ranging intellect, FM lectured, taught, and conversed on a future of post-nationalist politics, technological transcendence, and the optimistic future of humanity.

FM, the globetrotting son of an Iranian diplomat, saw all humanity as one community – a perspective later affirmed by the overview effect experienced by astronauts and cosmonauts. He felt that the technological offerings of the 20th century meant that we could redesign almost every possible human system, from parenting to education to industry. Many of his ideas clearly reflect the opportunities and perspectives of someone already in the cultural elite, and thus probably do less for the poor and marginalized that he might have thought. But despite this, he challenged those around him to develop a future-oriented and optimistic worldview. In that vision, he refused to acquiesce to our limitations, be they genetic, socially inherited, or baked into the process of cellular death. His chosen (and legal) name, FM-2030, represented both something technological (global radio communication) and temporal (the year he believed we would have solved the problem of death).

Throughout his career, FM argued that human beings have no right to be pessimistic about their fate. In his book, *Optimism One*, FM (then still going by his birth name, Fereidoun Esfandiary) decries the negative attitude that parades through human philosophies. FM saw religious beliefs rejecting the world and philosophical systems designed to help us accept the world as it is. But he believed that with

20th century technology we should be newly hopeful. We should plan on a bright future of abundance, and this message resonated with the budding community of transhumanism.

Among FM's coterie were Max More and Natasha Vita-More, both of whom would make vital contributions to the ongoing – if still somewhat fringe – transhumanist movement of the late 20th and early 21st century. While existential risk, the driver that would ultimately make transhumanism a close-to-household concept, had not yet emerged in our cultural conscious, More and Vita-More proved vital to maintaining transhumanist institutions, often finding ways to legitimate transhumanism through connections to academic and public communities.

From the outside, there appears to be something distinctly Californian about this entire enterprise. FM's group existed parallel to the Whole Earth Network, the broad California movement favoring anti-establishment communes, and counterculture intelligentsia as represented by Timothy Leary, the Esalen Institute, and similar organizations. A technofusion of counterculture, intellectual elitism, and a hyperbolic interest in human bodily perfection seemed almost inevitable. This led to the rise of magazines like *Mondo 2000* and *Wired*, which could provide public cheerleading for that technophilia. FM's salon, counterculture organizations, and the personal computing revolution all participated in and contributed to the new worldview.

The growing tech elite in San Francisco and Silicon Valley found a parallel path, one driven by the meteoric rise of personal computing and all the ways that placing affordable computers in domestic life changed how everyone saw the world. Not least, personal computers meant that likeminded thinkers could find one another from across the country. A modem (which allowed computers to connect to one another over the land-based telephone system) gave technoenthusiasts in Austin, New York, or Alabama a chance to find their tribe and carry out online conversations over early bulletin board systems.

A group of transhumanists and technofuturists were swift to join this movement. As a PhD student at the University of Southern California and a member of FM's crowd, More founded the Extropy Institute with his friend, T.O. Morrow. The Institute developed the Principles of Extropy which, though later adjusted, codified an essential perspective for many transhumanists. While most commentators have focused on the relatively naïve, and unquestionably overly individualistic, political libertarianism of the principles, More's and Morrow's larger contribution was enthusiasm for "perpetual growth." Drawing on FM's optimism, More felt that all people had a right to choose a path of growth for themselves and should have the opportunity to exercise that right into the indefinite, ageless future.

The Extropian vision of individual choice and perpetual growth echoed Huxley's vision of fulfillment, but it lost sight of his belief that fulfillment need be both individual and collective. The Extropians believed that each individual – barring any threat to others – should have total control over their own future. It would have appeared narcissistic to Huxley, who showed ongoing concern for the interlaced whole of humanity and the biosphere.

In the case of genetic manipulation, perpetual growth means the right to understand and alter one's own genetic code. In *Man into Superman*, Ettinger had already suggested that human beings might adapt their bodies according to whim or necessity (including, of course, growing tails...that strangely recurring theme in bioenhancement tracing from Haldane's *Daedalus* to the present day), and the Extropian crowd emergent around More, Vita-More, and their Internet community enthusiastically endorsed the idea that humanity can and should go well beyond therapeutic technologies to those that augment humanity to new states of physical, mental, and emotional health.

Predictably, a backlash against transhumanist futurism arose alongside technologies such as cloning, stem cell research, and genetic engineering. Hyperbolically labeled "the world's most dangerous idea" by Francis Fukuyama, transhumanism ran into an assortment of opponents. These included environmentalists worried about the impact on nature, political conservatives worried about the dignity of the human and the role of the Christian god, and political liberals worried about social justice and the shared human future. This confluence of opposition had a natural effect on the funding environment for such technologies. But technological progress has a way of overcoming such problems, as when Emmanuelle Charpentier and Jennifer Doudna enormously improved the CRISPR-Cas9 technology in 2012.

There is legitimate cause for slow and careful engagement with technologies that can affect the genetic inheritance of any species. We don't always know the long-term consequences of our choices. Spraying DDT helped eliminate malaria in the United States, but it is also critically endangered bird species (including the national bird, the bald eagle) and is difficult to remove from the environment once there. If we change the DNA of a species, that change can have ramifications beyond our immediate intentions and could well prove irreversible.

And so researchers build social coalitions to solve problems with the improved technologies of CRISPR-Cas9. In short, the technology finds specific locations in DNA and then cuts it, which permits the inclusion of new DNA. At a minimum, such interventions can repair genetic diseases; but moving from laboratory success to the environment causes concern. For example, worry arose that a 2025 therapy for high cholesterol could impact the liver. But in that same year, a baby was saved from a metabolic disorder thanks to CRISPR techniques and, as noted by Heidi Ledford in *Nature*, new companies seek to go where governments will not. Medical practice just affects individuals, however, and some proposed DNA interventions would affect the broader ecosystem; and it is in such projects that scientists must find allies. For example, MIT scientists want to introduce lab-engineered mice on islands in Massachusetts to show that they can reduce or eliminate Lyme disease there. Mice are the primary vector for Lyme disease; so making them resistant would be helpful to human beings, who are at risk from ticks transferring the bacterium. The value of an island experiment is that it can show results in a semi-controlled environment, and it would be a prelude to widespread deployment of the modified mice.

The Lyme experiment requires a slow regulatory and public process, but there are genetic champions who don't consider waiting an option. Biohackers, for example, affirm their right to biological self-determination. Many refrain from typical transhumanist commitments, such as the pursuit of immortality. But they overlap with other transhumanists in their commitment to self-exploration and individual freedom. Some, named "Grinders" by scholar Jacob Boss, live in the shadow of institutionalized transhumanism. Unlike the AI entrepreneurs in Silicon Valley or the roboticists at Carnegie Mellon and MIT, the grinders operate out of homes and garages, and, occasionally, rented community venues. They reject interference from governments and industry, experimenting on themselves so that, as Boss tells it, "no one is denied access to life-enhancing technologies."

The technological benefits to human genetic engineering are too strong for political or economic headwinds. There is no reason, in principle, to believe that human DNA cannot be altered to eliminate disease; and so people persist in research that will surely have benefits beyond that point. For transhumanists, progress in genetic technology promises an end to limits imposed on us by our biology, and it is important to keep in mind that they have ready allies in anyone affected by genetic disease. There are many people who suffer from illnesses and conditions that we may overcome through genetic engineering.

My own experience as a teacher reveals an ongoing shift in young people's perceptions. Early in the 2000s, when I gave my students an anonymous poll on whether they would genetically engineer their children, only a few would say yes – usually just two or three out of a class around 25. The rest quickly acknowledged, however, that their own children might choose otherwise: it takes only a few obviously enhanced people in the community to make the technology more appealing! As I write this in 2024, it is already the case that one-third to one-half of the students I poll desire genetic enhancement, and that feeling is nearly universal among those with chronic health conditions.

If our biosciences pursue genetic engineering, we will be wise to guide it as best we can. Many religious voices speak out against genetic engineering as "playing God," but others see the need to engage with scientists and policymakers. Christian theologians like Ronald Cole-Turner and Ted Peters argue that if their god did not want human beings to develop such powers then we would not be able to. The judicious use of those powers, however, is up to us. Similarly, in Judaism there is a longstanding principle that the saving of lives trumps all other considerations. From such religious perspectives, it becomes a necessity to undertake research that can accomplish those goals. But it simultaneously requires that human beings undertake research with restraint and with safety in mind.

Not all transhumanists believe in restraint or believe that we have time for it. Genetically modifying people to adapt to climate change might become necessary, for example, as access to food, clean air, or comfortable temperatures becomes scarce. In fact, for some transhumanists, preventing these kinds of threats is a key opportunity in genetic engineering. Remember that threat, especially existential threat, is key to the entire enterprise of futureproofing humanity. In this sense, the clear and present dangers that we face as a species become the drivers to an interlocked set of technological practices.

The evolution of humanity in the face of existential threats would require a response at the level of DNA. It is unlikely that human beings could remain limited by our contemporary biology should we spread into the solar system and beyond. In fact, the futureproofing religion coheres around the idea of human beings inhabiting other worlds. Interstellar humanity will be the subject of chapter five; but for the time being it is worth noting that human beings have evolved to live on Earth, not on the moons of Jupiter. Genetic engineering could lower our oxygen needs, adjust our bones and muscles for different gravity, or provide us with alternate energy sources.

The connection between spaceflight and human evolution has, perhaps, no more influential driver than the 1968 film, *2001: A Space*

Odyssey. In that film, and the accompanying novel by A.C. Clarke, a human being follows alien technology "beyond the infinite," where he evolves into the "Star Child." Envisioned in the film as a gigantic baby floating through space, in the novel, the Star Child defuses the world's nuclear arsenal using inhuman powers of the intellect. Upon its return, Clarke writes, the Star Child is "master of the world." Bewildering as this end may seem, for two generations of late 20th century thinkers, *2001* inspired faith in a new evolutionary future, one that surpasses the confines of present-day human biology.

Whatever our distant biological future might be, the religion of futureproofing demands activity now. We will not be prepared for interstellar life if we do not find solutions to our current conflicts and threats, and we will not find those solutions – it is supposed – without technological advancement. The careful reader will, of course, note that many existential threats are outcomes of human activities. From nuclear war to genetically enhanced pathogens to climate change, human beings have made many choices that produce the very risks we now wish to invent our way past. Fortunately, we bear no responsibility for other threats, such as supervolcanoes or giant asteroids. But whether we attempt to invent our way past cosmic accidents or human folly, the mere risk of such dangers suggests the need to act decisively.

Market forces and unearthly DNA

Unfortunately, to whatever extent we make choices about our future, we are driven by economic and political structures that supersede individual and even our collective interests. When government regulators are former industry lobbyists (and vice-versa), it is difficult to ensure that public need competes effectively with the needs of corporate profit.

In response to economic risk, then, some transhumanists vigorously pursue outside-of-the-box solutions to genetic engineering. Among such "biohackers," Josie Zayner is particularly noteworthy. She caused a stir when she publicly used a CRISPR injection targeting the myostatin gene, a process she had been using to improve muscle mass in frogs. Beyond this, Zayner operates a business shipping CRISPR-Cas9 experiments to individual consumers. For approximately $200, a buyer can purchase everything necessary to conduct a small-scale experiment.

Unlike nuclear bombs (which require rare elements and massive labworks), biotechnology can be conducted at small scale and minimal cost. This is one of the reasons that genetic engineering represents an existential risk. After all, it takes only a modest laboratory for someone with a dangerous ideology to try and manufacture a highly contagious and/or lethal disease. Unlike nuclear weaponry, it does not take government action to advance research. So Zayner provides convenience and instruction but is not herself creating any risks that don't already exist.

What Zayner and other biohackers do, however, is reject the social and political environment in which biotechnology gets conducted and resist corporate control of biotechnology. While large corporations are not an existential risk, they make profit generation the key value of biotechnology and genetic engineering. Zayner thinks of herself, perhaps rightly, as democratizing genetic engineering. From the biohacking perspective, people have an individual right (per the Extropian movement) and a political need to share in the technologies of the future.

Should the biohackers circumvent corporate control, they might help humanity avoid the kinds of risks that science fiction has described for years. Much of dystopian science fiction is not really about the technologies or even their outcomes (such as pollution); it is really about the political and economic structures in which those technologies emerge. At a minimum, if genetic technologies remain exclusively in the hands of the wealthy and powerful, then new forms of oppression and exclusion will emerge. The film *Gattaca* points toward our existing forms of social inequality and economic dispossession but argues that genetic engineering will worsen them both. In a world where better genes can be bought, the "gen-poor" will lose opportunities in work, leisure, and love.

Thus, the enthusiasm that drives the futureproofing agenda has many prongs. It responds to future dangers at the same time as responding to dangers in the here and now. Simultaneously, it poses risks of its own. Whether it offers the right solution to any of the dangers is, of course, open to reasonable debate.

Plausibly, even a successful biohacking revolution and/or the deployment of widespread genetic engineering could leave humanity less happy than before. Regarding this, neuroscientist David Eagleman points to a "paradox of choice" that would apply to bioengineered children. The paradox states that the more choices a person has, the less likely that person is to be happy with the choice they make. That is, if I am admitted to 100 universities and select one, I am less likely to be happy than if I had been admitted to 3 universities and selected the exact same school. In these matters of choice, it is not the chosen situation that dictates our happiness but our imagination of all the things we did not choose. If this is, in fact, legitimately applicable to human engineering it raises doubt about our happiness when we design ourselves or our children. Suddenly we can easily prefer an imaginary child to the one we do have.

But there is no question that genetic engineering offers substantive hope for humanity. At its most basic level, it includes fending off disease and overcoming crippling disabilities. The MIT mice would have to be engineered to create a solution that is really about other species (the ticks and the bacteria). But human beings carry genetic anomalies that cause direct hardship. A genetic solution to Tay-Sachs disease, for example, would have nothing but beneficial effects, creating a cure for a disease otherwise universally fatal in childhood. Since the cause of Tay-Sachs is a faulty genetic structure, fixing it comes with no risk of adverse effects as long as nothing else gets changed.

These kinds of solutions, some confined to human beings and some with their primary impact on other species, are the stepping-stones to a world of genetic augmentation and one element in the futureproofing project. What solves problems today opens the door to greater intervention tomorrow. If we can cure Tay-Sachs, can we

59

cure aging? The idea that aging is a disease has gained steam in the 21st century. Not only is aging specifically correlated with a host of other diseases that might go away if we stopped getting older, but aging is, itself, subject to intervention.

Alien aficionados might like us to believe that our current genetic inheritance is the deliberate action of extraterrestrial intelligence; but it is rather more likely that we evolved through chance encounters over billions of years. Though we probably don't have our origins in space, if we are to move there in the future we will need DNA recoding that targets our new living conditions.

Human beings evolved to conditions on Earth, with limited geographic variation and specific attention to other forms of life. In short, we enjoy mountains, lakes, and trees. If we live away from Earth, we will either need to terraform those other places (i.e., make them like Earth) or reprogram ourselves to match the existing conditions. There will be little point living into the future, whether on Earth or beyond, if we are not able to be happy. So spacefaring earthlings will almost certainly bear different genetics than do we.

Getting beyond the Earth could necessitate changes like efficiency with air, food, and water consumption to make space travel reasonable. But creating efficient human beings won't make for happy human beings. And so, as humanity considers life beyond our atmosphere, it behooves us to contemplate how we can ensure physical and emotional health in such a place. It may not be enough to isolate "contentment genes" in the happiest human beings; the ability to be content on Earth is no guarantee of contentment on Mars or the moons of Jupiter or during months-long spaceflights. While astronauts have managed more than 12 months in the International Space Station, that is with the knowledge that they *could*, in principle, come back to the surface. Anyone on a journey to the moons of Jupiter is on that journey until its completion. And so genetic changes would enable the very future that necessitates them.

Genetic changes to make humanity comfortable in space may never transform us into cosmic star children with telekinetic powers. But they seem likely to produce a new biological species. In the event of humanity genetically changing the species, we might need to rede-fine what makes us human. Are our genes critical components of our humanity? If so, which ones? If not, then what *is* essential to being human?

Species saved

Our world comes with a variety of problems, and some of these look existentially dangerous. The viable solutions, say transhuman-ists, include rapid investment in genetic engineering. We could have widely distributed individual technologies such as those purveyed by Josie Zayner or we could have corporatized child development. Either one might enable our flight from Earth or deliver an escape from the problems facing us.

Otherwise, solving Earth's problems could require an engineering effort that goes well beyond just one species. Every year we see species dying out, arable land declining, and weather patterns shifting (almost always for the worse). One major effort that would probably look quite attractive to Julian Huxley is underway in Russian Siberia: Pleistocene Park. It is here that Sergei Zimov and his family hope to stave off climate disaster by rewilding the Siberian steppe and resurrecting its past ecosystem. Key to this challenge is restoring large herbivores to the region. When human beings drove woolly mammoths extinct, we inadvertently removed a crucial part of the grassland's lifecycle. In the absence of large herbivores, the grass-lands failed and tundra took over. Thanks to Zimov's effort, there are now reindeer, horses, bison, oxen (unfortunately all of whom turned out to be male), and other species of large herbivores. If Harvard geneticist George Church can succeed in replacing some of the DNA in elephant embryos, even "mammoths" will return. Should a viable arctic grassland return, humanity will gain leverage against Earth's warming temperatures.

The thawing of arctic permafrost is one of the greatest potential drivers to runaway climate change. As the permafrost melts, methane gets released. The methane has been down underground for millennia, unable to escape from the frozen landscape. But today, there are lakes where methane bubbles upward...making it possible to start a fire on the surface of the water. As global temperatures continue to rise, more methane will escape, driving more rising temperatures. The extent of the danger is not fully known, but the reality of it is unquestionable.

In *Pleistocene Park: Extinction and Eternity in the Russian Arctic*, Anya Bernstein tracks the unfolding of Siberia's methane cataclysm and Sergei Zimov's scientific effort to prevent it. In order to keep the methane safely underground, Zimov wants to rebuild the tundra that existed tens of thousands of years ago in the Pleistocene Era. Environmental icon Stewart Brand, founder of the *Whole Earth Catalog*, supports the project and Church's effort to "resurrect" the mammoths using ancient DNA and modern elephant eggs.

This movement is a direct response to existential risk. As Bernstein writes, "Pleistocene Park is *survivalist* not in the usual sense of preppers seeking to ensure their individual survival in the case of a global catastrophe but as a project aimed to prevent the extinction of the whole of humanity" (emphasis hers). Scientists like Zimov believe that, with the right push, animals will do the heavy lifting in ecological restoration and climate change mitigation; but it's obvious that the entirety of this project is one of human innovation and architectural design. Human beings may have driven the Siberian megafauna extinct but getting the landscape ready for them and then taking bison on a three-week sailing adventure across the White Sea is hardly a job that nature can handle without a human hand. That task sits right in the core of biotechnological futureproofing. As Bernstein puts it, the park is "an anthropocentric project undertaken to benefit humans, specifically, to enable human survival into the future."

The bioengineering of Pleistocene Park is not new to humanity. The prairies of the midwestern United States, for example, were long maintained by Native American communities through controlled burns. They witnessed lightning bring prairie fires and the subsequent growth of grasslands lured bison to graze. They then spent millennia curating the grasslands, using them for medicinal plants, food, and to create hunting grounds. My own institution, Knox College, uses controlled burns to this day, as it is the only way to promote the growth of the prairies in our Green Oaks Biological Field Station. It is our own contribution to this cycle of death and life, ash and growth, that prompted my collaboration with artist Michael Takeo Magruder on his *re:Generated Prairie* exhibition at Knox College. In our work together and, especially, his photography, videography, and subsequent digital reconfigurations, we revel in the human encounter with nature: our vision is always technologically mediated, and we always make crucial contributions to the environment we occupy.

While ecosystem management has a long history in human practice, the Pleistocene Park project is a form of revolutionary science, a shift away from old paradigms in multiple disciplines. The idea of scientific paradigms traces to Thomas Kuhn's famous and brilliant (and thus often maligned) *The Structure of Scientific Revolutions*. Kuhn argues that most of the time science just plods along solving relatively mundane problems until too many unexplained and unexplainable anomalies add up. Then some folks experience a crisis that provokes a shift from the old paradigm to a new one. Kuhn's is not a perfect description of the history of science (hence the criticisms), but it's an awfully good one (hence the people who want to criticize it). Zimov's project challenges our understanding of the climate, the ecosystem, and the role of humanity within it. Zimov sees us as degenerate scavengers but also suggests that we can move (backward?) to a different approach to environmental management. We're part of it all, and we can actively design ecosystems without degrading the natural world.

It is the simultaneity of resurrection, restoration, and preservation that would appeal to Mr. Huxley. As the first director of the United Nations Educational, Scientific, and Cultural Organization (UNESCO), Huxley committed much of his professional life to the preservation of cultural and ecological resources. The potential of projects like Pleistocene Park to not only bring back resources once lost but to maintain continuity into the future represent the kind of cosmic responsibility that Huxley suggests. Building Pleistocene Park seems low-risk, but there are also more dubious efforts at planet-wide solutions: geoengineering projects that don't quite fit in the mythical story of futureproofing. Perhaps it is their very lack of magic (and not just their cataclysmic risk) that makes things like gigantic mirrors in space look dubious. Pleistocene Park offers a return to Eden and a resurrection of ecosystems and even species. It thus fits in the bioengineered world of futureproofing.

But the commitment of Pleistocene Park is to ensure an environment habitable by human beings as we are; most futureproofers are not satisfied that this will answer. Instead, they want to modify humanity. The biotech enhanced lifestyles that we discussed above are higher risk but also higher reward. They promise solutions to a host of problems and open new vistas to the human imagination. Resurrecting mammoths will be wondrous, but it won't help us live on Mars. It might not even go far enough to help us live on a post-climate change Earth.

Geneticist George Church believes, however, that these technological advances will provide the steppingstones to spaceflight. In a 2012 book co-authored with Ed Regis (author of the first ever essay on the Extropians, published in 1994), Church writes "supposing that one day we have managed to conquer an impressive array of human maladies, including infectious disease of almost every kind, we might then turn our attention to a future era in which members of the human species might occupy an entirely new ecological niche...on another planet. Mars, for example." He is not alone: this ethos permeates *The Next 500 Years* by geneticist Christopher Mason, who notes that we "are overdue for a planetary-wide risk that threatens not only us but all life on Earth" and thus that "the same innate, biological capacities of ingenuity and creation that have enabled humans to build rockets to reach other planets will also be needed for designing and engineering the organisms that will sustainably inhabit those planets."

Rolling (and re-rolling) the genetic dice

I imagine that, more or less like zombies, the biotechnological promise of eternal youth and genetic enhancement are beyond anyone's ability to kill. Their resurrection in 21st century science, technology, and venture capital may or may not yield fruit, but it won't be the last we see of them either way. Biological enhancement is part of the human story, or the story we tell about humanity. Barring an actual cataclysm, we will continue to see transcendent promises cycle through hype cycles. And at some point, they may even come to pass.

When we look at how such sciences fit in our larger storytelling landscape, we see an attempt to reconcile those future possibilities with present realities. The promises of geneticproofing are actually rather narrow; they come from authors wearing blinders, and who thus see only one possible future path of (probably elitist) redemption. But the stories we experience in literature, film, and games encompass that path while suggesting alternatives.

One of the great things about games, in particular, is how they reject any possibility of a linear, predetermined future. After all, quality game design automatically balances existing conditions, player decisions, and an added element of random chance. A player lives in a world, most likely a world created by some other force (just as people currently live in a world designed by evolution and geology, or perhaps by gods, or perhaps by some combination of those). And when players make choices in their games, the outcomes often depend on the roll of a die or the spin of a wheel.

Stories are predetermined but games are open ended. Francis Bacon told the story of a biotechnologically transcendent Christianity...one he could believe to be inevitable by the grace of his god. Geneticists like Church and Mason likewise tell a story of biologically-enhanced humanity (with regrown tails, of course), a story that seems all but inevitable just as long as we don't drive ourselves extinct before we get there. Whether implicitly or explicitly, these dreams of technological transcendence carry the air of inevitability. In doing so, they restrict our vision. Games, however, open our eyes to new opportunities, better recognize the potential for catastrophic failure, and give us a chance to explore.

I remember playing *Shadowrun* in the 1990s, reveling in the power that my cyborg characters gained through technological enhancement: cybereyes to see in the dark, targeting systems and enhanced reflexes to help in combat, and a brain-computer interface that allowed me instant access to new skills – perhaps the inspiration for when Neo unplugs from his training simulation and calmly asserts "I know kung-fu."

Shadowrun is a cyberpunk science fiction roleplaying game in the tabletop genre pioneered by *Dungeons & Dragons*. By drawing on your character's skills, equipment, and "cyberware," you travel through a science fiction world, led by the game master who describes what happens and adjudicates the rules. A player rolls the dice and seeks to combine good decisions with good luck. If all goes well, the character lives to see another day, thriving in the midst of chaos and danger.

Today, *Shadowrun* strikes unfortunately close to home: corporations act like nations, individuals exist in the margins between them, and the environment is in a state of near-collapse. Looking at Tesla's Cybertruck, old school players instantly recognize the aesthetic of *Shadowrun's* vehicles, especially the sharp edges and geometrical designs present in the *Rigger's Handbook*. Elon Musk and his associates want to build the car, the corporate oligarchy, and, supposedly, the cure to all of life's ills. In a world where technological fairytales risk becoming ghost stories, it's no wonder that we can see the influence of imaginative games.

The compelling side of games like *Shadowrun*, however, is that they reveal cracks in the dystopian firmament. There is a place for the players to run and hide and carve out a niche of their own. And since the game does not lead to a foregone conclusion, even hard to imagine futures are, despite their improbability, part of the landscape of possibility. Perhaps someday we will have enhanced reflexes, superhuman muscles, and brain-computer interfaces, perhaps we will have genetic interventions that liberate us from poison, sickness, and death.

Where *Shadowrun* leaves off (apparently at low-earth orbit), the game *Eclipse Phase* launches into a transhumanist smorgasbord of alien, human, "uplifted" animal, and machine lifeforms. In doing so, the game proposes that we may in fact transcend our current reality. In *Eclipse Phase*, life is without limit: at death or in search of a light-speed flight across the solar system, a digitally transcribed mind rises up in a new body. The entrants to transhumanity no longer worry about disease or old age. Perhaps our life among the stars will open up all the genetic opportunities that Mason promises in *The Next 500 Years*. In *Eclipse Phase*, also a tabletop roleplaying game, players create their spacefaring characters with a scope that appropriately revels in radical new bodies.

The mantra of the 2019 second edition of *Eclipse Phase* is "Your mind is software. Program it. Your body is a shell. Change it. Death is a disease. Cure it. Extinction approaches. Fight it." Built around the very conflicts over climate change, wealth inequality, partisanship, and AI that drive our fear of existential risk, *Eclipse Phase* allows players to head into the solar system and beyond, seeking survival and new possibilities. The transhumanist recomposition of human bodies, in fact their rejection as temporary and transformable, explores the biological pursuit of renewed and redemptive life. Even in a dystopian future, there exist pockets of opportunity, and the players of *Eclipse Phase* rethink the meaning of life in a world where that commodity extends beyond our present expectations. In the fantastic universe of *Eclipse Phase*, life over-flows out of biology, and futureproofers follow in that trajectory. Perhaps reengineering our DNA and cells won't be good enough and we might need to shift to ever more revolutionary interventions. Perhaps games like *Eclipse Phase* help us figure out how to live in a bioengineered future.

Going backward, I'll note that it's possible our innate desire to play games could be the origin of religion itself. That's what Johann Huizinga argued in the mid-20th century, and it is the starting point for many scholars seeking to understand games, especially videogames and roleplaying games. When we see religion, science, and technology tied up together in futureproofing it's no wonder that there are games to unravel the knots and help us sort matters out. Huizinga surely couldn't have predicted the geneticproofing movement in the 1940s, but he wouldn't be surprised that we can play our way to a better understanding of it.

In futureproofing, genetic and biotechnological interventions suggest the first opportunity to extend human lifespans, human capacities, and the environments where (post?)human beings can live. Immortality and perfect health are the hallmarks of human salvation, at least in the western traditions. Life in the stars even accompanies their incarnations in biotechnology, offering a mirror to traditional religious dreams.

For much of the 20th and early 21st centuries, biology suggested the most profound knowledge and offered the most profound opportunities. To study biology was to study ourselves and search for transcendence; there were scientists for whom biology was the final frontier. But for many transhumanists, it was a final frontier of a different sort. Supposing that biology will be forever subject to limits, futureproofers also draw on advancements in computer science to seek an immortal afterlife.

CHAPTER THREE
CYBERNETIC ANGELS

In the last chapter we read about how existential risk prompts enthusiasm for biotechnological interventions. The futurists of biotech believe that engineering humanity is necessary for three reasons: to overcome our limits, to compensate for problems incurred through our other technologies, and to expand the biological options for humanity beyond Earth. While many futurists believe that these are necessary technologies, the most ardent supporters of future-proofing humanity feel bioengineering will prove ultimately inadequate to the task. That is, genetic engineering might earn us a few more years, but it cannot provide sufficient velocity for humanity to escape its impending doom. For these futurists, the merger of biology and technology, and the transformation of the former into the latter, is necessary for our species' survival into the indefinite future.

According to the unwritten futureproofing manifesto, the future belongs to the machines. According to Elon Musk and others, human beings are likely "the boot loader for superintelligent AI." This analogy refers to the small(ish) program stored in a read-only format in a computer, which tells the computer how to access the machine hardware (e.g., the hard drive or other storage medium) and get programs like the operating system running in the computer's memory. After doing so, the boot loader is irrelevant until the computer is shut down and restarted. While Musk supposedly fears the replacement of human beings by machines (even while building a large language model AI called Grok), others relish that prospect. To maintain relevance, they say, humanity will have to transcend DNA altogether, we will bind our flesh to steel and perhaps even replace our bodies entirely.

According to futurists of all stripes, AI brings significant risks: if the future belongs to machines, then those machines imply inherent risks for humanity. The risks of AI are manifold. The most likely risks, of course, are surveillance culture (which we already have), deepfake scams (again, which we already have), hacking or system failure of autonomous weapon systems (yup, we have those also, and they can choose their own targets), and the further entrenchment of human bias in supposedly neutral systems (yes, we have that too). But other futurists look toward even scarier outcomes: there are Terminator scenarios where machines reject our authority and try to kill us and there are absurd claims about AI turning all of us into paperclips just because they are designed to "maximize" paperclip manufacture. Terminators are unlikely and the paperclip "thought experiment" is silly, but more plausible dangers – like the increasingly militarized universe of AI – accompany these. From direct and obvious risks like unemployment to more speculative dangers like Skynet, we should not discount the variety of threats posed by AI.

In the world of futurism all those risks add up, and this compels worry among AI practitioners. While coding away to build AI products, people in the industry fear AI could become an existential risk. According to a survey of 841 engineers conducted by the consulting firm Amplify Partners in 2023, an alarmingly significant number believe that AI will threaten humanity. *Averaging across the respondents, AI engineers see a roughly 40% chance that AI will destroy the world.* Given the proclivity to blackmailing shown by Anthropic's AI, Claude, and evidence that GPT can subvert instructions to shut down, perhaps by 2025 engineers would already revise that number upward. The percentage chance of AI destroying humanity, colloquially known as p(doom), seems to have become a common topic of conversation.

This fear isn't just about the machines; it's also about the people building the machines. Such distrust has been readily visible since the launch of ChatGPT. At one point, the Executive Board of OpenAI fired its CEO, Sam Altman, for being opaque to the point of dishonesty. A strange revolution took place, one that resulted in the board members' ouster and Altman's reinstatement. Shortly after, however, Altman's antics threw more of his fans to the winds. This was dramatically revealed when OpenAI seemingly used a facsimile of the actress Scarlett Johansson for their ChatGPT4o voice system (despite her refusing them permission). On a forum dedicated to effective altruism, psychologist Geoffrey Miller noted his belief that Sam Altman was showing himself to be deceptive and arrogant "with no serious concern about the extinction risks he's imposing on us all." By 2025, when Open AI started campaigning for government bailouts and then claimed they weren't, most neutral observers found it hard to believe Altman's disingenuous backtracking.

In essence, it's not just a matter of whether intelligent machines will endanger us, it's that our corporate leaders relentlessly pursue their own goals regardless of the potential impact they impose. The dissolution of the AI safety team at OpenAI along with the elimination of similar groups at Microsoft, Google, and other tech companies adds fuel to this fire. For these and other reasons, some AI researchers have spoken out to encourage regulation and public engagement, as when a number of OpenAI employees signed an open letter calling for better oversight and protection for whistleblowers. Meanwhile, OpenAI moves forward with an effort to finally make money on their very expensive, low revenue product: by launching erotic chatbots and by constantly surveilling users through pendants equipped with cameras, microphones, and access to GPT.

Fear of AI goes back decades. Of course, there is Mary Shelley's monster, driven mad by the rejection and isolation inflicted by Frankenstein and the rest of humanity. In his seminal play that coined the term "robot," *R.U.R.* (first performed in 1921), Karel Čapek recognized the threat of robots but he also knew that danger actually results from the violent tendencies of human beings. Many commentators have noted that the book ends with humanity driven extinct and robots acquiring the powers of love, empathy, and, it seems, even the power to reproduce. What no one seems to note, however, is that the robot uprising happens because robots are forced into combat by their human masters – this fact is more important than how they gain free will. Were human beings less inclined to build weapons of war and rule over one another, they might have remained quite safe in *R.U.R.*

As technology developed, so did the storytelling; but by the late 20th century, the stories were in popular science, not just science fiction. The cover illustration of Kevin Warwick's *March of the Machines*, for example, showed armed robots marching in military formation. In response to the risks we face, he advocated cyborg technologies that help people advance along with the robots. Curiously, their own fear of robots replacing us, long apparent in science fiction, drives some scientists to compose pop science books that urge more, not less, investment in those technologies.

The futureproofing agenda for digital technologies begins with cyborg enhancements, but it ends in the complete transfer of human consciousness from our biological bodies into machine bodies. There are scientists in robotics and AI who believe that cyborg enhancement is the end goal, but the majority of futureproofers think we will need to go further to meet our ultimate evolutionary potential.

The technologies of AI and robotics are themselves supposed to be existential threats that we can overcome only by becoming, ourselves, robots of one sort or another. As the saying goes: if you can't beat 'em, join 'em. Robotics and AI will also be key to the space-faring posthumanity that we keep dancing around, and which we will articulate fully in chapter five. For the time being, it is enough to reveal how "competition" between human beings and machines drives the belief that human salvation is postbiological, that we must adopt an AIproofing mindset.

The intelligence explosion

In the 1960s, mathematician Irving Good predicted an explosion of machine intelligence, expecting that progress in computers would lead to "ultraintelligent machines." This perspective circulated in the computing world, especially since the onset of the 21st century. Under the assumption that technological progress undergoes continuous acceleration, many advocates of computational superintelligence argue that machines will soon equal human intelligence, and, once that happens, AI will almost immediately far surpass human ability, thus bringing about the end of human dominance.

Much of the faith in accelerating progress is based on Gordon Moore's 1965 observation that we were doubling the number of transistors on an integrated circuit every year, thus effectively doubling the speed of computers in that timespan. There are limits to how fast we can make an integrated circuit by shrinking the transistors; and more recent advocates have broadened Moore's Law to include other computational upgrades, arguing that the doubling of computational speeds will persist indefinitely through as yet unthought of technologies. Although later revised to claim that computers double in speed every 18-24 months, Moore's Law provides the bedrock for the claim that computers will outstrip human intelligence.

For Moore's Law to really mean that computers will outsmart us, we must pretend that computational speed is the same thing as intelligence. And that's a rather poor thing to pretend. It is profoundly unlikely that simply speeding up computers would turn them intelligent or conscious. I suspect the adoration for "scaling" AI training in the GPT-era implicitly recognizes that speed doesn't amount to intelligence. Of course, neither does scaling (unless you have a financial stake in getting another venture capital infusion for your profitless company). Deep down, advocates of the intelligence explosion know this, so they present broader arguments about technological development. In 1983, mathematician and science fiction author Vernor Vinge wrote that a favorite game of futurists is to plot technological performance – computer speed, say – against time. Such trend curves climb ever more steeply. Extrapolated 30 or 40 years they are so high and steep that even the most naïve futurist discounts their accuracy. Some who talk about that era predict a leveling off of progress. After all, saturation effects are observed in other processes. There is an important reason why this process won't level off. We are the point of accelerating the evolution of intelligence itself...We will soon create intelligences greater than our own. When this happens, human history will have reached a kind of singularity.

Vinge was part of a movement in its infancy, one drawing on science fiction but also on the claims of scientists like MIT's Marvin Minsky and Carnegie Mellon's Hans Moravec (whose first publication on the subject in appeared in 1979). Collectively, as Vinge eloquently predicted, the future-oriented thinkers saw a kind of cosmic necessity in the arrival of machine intelligence.

This argument is diffusely present in the work of roboticist Hans Moravec. In his book, *Robot: Mere Machine to Transcendent Mind*, he argues that evolution is "weeding out ineffective forms of thought" and thus our replacement by machines is inevitable. For Moravec, there is some form of Darwinian competition between human beings and machines – though what, precisely, is the ecological niche over which we compete is unclear – and that competition guarantees the ultimate superiority of machine intelligence. He takes the rapid improvement in computer vision to prove that the improvement in computers is not just a matter of computational speed but also indicates a more general phenomenon of computational progress.

Drawing on Moravec but going farther, AI innovator Ray Kurzweil claims that the entire universe is subject to a "law of accelerating returns." He believes that any and all "ordered systems" will increase their own order exponentially. The entire universe is subject to entropy, the increase of chaos. But ordered systems draw into themselves: while the universe outside an ordered system becomes more chaotic, within the boundaries of the system order increases.

And since the Earth is fundamentally orderly, drawing energy from the sun to resist the chaotic entropy of the universe beyond, everything on Earth increases its order. As he details it in *The Singularity is Near*, all manner of technological systems on Earth will inevitably increase their own orderliness...and at an accelerating pace. So, the amount of order gained over a specific period of time (let's pretend it's one year) will double over the next iteration of that time period (the next year). He claims that "the nature of time is that it inherently moves in an exponential fashion" and connects this to technology, which "picks right up with the exponentially quickening pace of evolution."

Kurzweil believes that Moore's Law is just a specific example of the broader law of accelerating returns. The systems of economics, industry, and technical expertise that enable the creation of integrated circuits – and computers more broadly – are orderly systems. Kurzweil thus claims that they accelerate their own development of order. And so, the technological returns on 18-24 months of time will double every repetition of that time period. Barring a total collapse of human civilization, this would mean that computers will *always* double in their speed and capability every 18-24 months. But it also means that every accompanying technological and scientific process will also be doubling its output on regular intervals, though the time period for any given technology will vary.

Doubling the output over specific timeframes (like the 1-2 years of Moore's Law) produces exponential growth, and if that process continues it would result in what has been labeled the Singularity. If the "intelligence" of computers doubles over and over again, eventually it will equal human intelligence. Then, just a couple years later, the computers would be twice as intelligent as people, then four times as intelligent, then eight times, etc. The Singularity is that moment in this technological growth where change happens so fast that we literally cannot predict what will happen on the other side of the next doubling. Imagine a moment where robots are eight times more intelligent and capable that we. It's hard to imagine, but roughly speaking we might know what that looks like. What about when the machines are sixteen times more intelligent? Do we know what that looks like? What about from thirty-two times as intelligent to sixty-four times as intelligent? At some point, this exponential progress yields a world so radically different from our own that we literally cannot conceive it from our vantage. The Singularity is that radical break in history: it would be a moment of cosmic significance, not your typical human revolution but a revolution in humanity and life itself.

If, and this is a big if, technological systems continue doubling their potential at regular intervals, computers will inevitably exceed humanity. According to Moravec, Kurzweil, and their followers, a computer should equal human intelligence in the first half of the 21st century. Kurzweil's more aggressive predictions have already failed to hold up, but he remains committed to his position with slight revision. Moravec's date for a human level computer is 2030; so, he remains as yet uncontradicted by reality... but his predictions will soon be tested.

In the face of reality, Kurzweil retconned the law of accelerating returns in his book, *The Singularity is Nearer*. Published in 2025, by which time many of Kurzweil's predictions remained unfulfilled, the book moves the LAR (rebranded as LOAR) from being "a basic attribute of time and chaos" (in *The Age of Spiritual Machines*) to something that describes "information technologies like computing." He further reconfigures the LAR by arguing "my work has sometimes been mischaracterized as claiming that technological change itself is inherently exponential, and that the law of accelerating returns applies to all forms of innovation. That's not my view. Rather, the LOAR describes a phenomenon wherein certain kinds of technologies create feedback loops that accelerate innovation." This is a far cry from when he previously argued that "the nature of time is that it inherently moves in an exponential fashion."

Of course, the LOAR-as-exclusive-to-information-tech position contradicts the specific claims he made in *The Age of Spiritual Machines*. There, he argues that "the other significant point is that technology, like the evolution of life-forms that spawned it, is inherently an accelerating process." In that earlier book, he describes the acceleration of technological development with regard to canals, paved roads, and bicycles along with lightbulbs, motion pictures, and telephones. At that time, he argued that "all the fundamental processes we have examined – the development of the Universe, the evolution of life-forms, the subsequent evolution of technology – have all progressed in an exponential fashion." A few years later, in *The Singularity is Near*, Kurzweil again positions technological development as broadly accelerating: "the ongoing acceleration of technology is the implication and inevitable result of what I call the law of accelerating returns, which describes the acceleration of the pace of and the exponential growth of the products of an evolutionary process. These products *include*, in particular, information-bearing technologies such as computation" (my emphasis on "include"). His vision of accelerating returns was, obviously, much more sweeping at that time.

Regardless of how Kurzweil defines the LAR, he sees human-equivalent machines on the immediate horizon. In *The Singularity is Nearer*, Kurzweil claims that by 2029 there will be a neural net computing system that is better than all people at all tasks. Soon after, we would find ourselves debating the consciousness of such machines.

Whether a computer has reached human level will be exceedingly hard to decide, but someday it may be necessary to do so. The most famous attempt to define the constraints was proposed by Alan Turing, whose 1950 paper, "Computing Machinery and Intelligence" puts forward the now famous proposal that we just make our decisions on the basis of whether or not the computer can fool us into believing it's a human being with sufficient frequency. Although he rejected the essential idea of worrying about whether machines can think, he felt that by the year 2000 people would routinely speak about that being the case "without expecting to be contradicted."

As in technological forecasting more broadly, Turing rather missed the mark with that date. But recent improvements in large language model chatbots have lots of people claiming that AI can think, can feel, can be a person. Research by Clara Colombatto and Stephen Fleming indicates that a stunning 67% of those surveyed believe that GPT experiences some form of consciousness, and the more frequently a person uses GPT the more likely they will feel it to be conscious. The Turing Test has lost its luster, but the problem of machine consciousness definitely remains.

I feel very comfortable saying that, at this point, no computer is even a little bit conscious; but I don't feel equally comfortable saying it could never happen. If we're going to witness increasingly sophisticated AI behaviors and communication, at some point we need to sort out whether it is more than just a blind programming or prediction machine and whether it actually understands the world and its role within it.

As we'll see, plenty of AI engineers and users already think there are conscious machines, but they do so because the contours of their work promote precisely that perspective. The famed android designer Hiroshi Ishiguro once told me that with a careful choice of context robots become indistinguishable from human beings. This matter of context is key. Early in the 2000s, he deployed his "geminoid" as a receptionist and believed nearly 80% of people who interacted with it failed to recognize it was not human. This seems to indicate two things: 1) many people do not notice if something is a robot when they don't expect it to be, and 2) even very brilliant people who design robots overstate the likelihood of people doing so. Fundamentally, there is very little chance that 80% of people could not distinguish between the android and a human being in the year 2010, though certainly some people must have failed to see the difference. These same dynamics no doubt help us understand users' and designers' beliefs about AI technologies like GPT. These dynamics force us to think about what conditions would mean that the vast majority of people cannot distinguish between human beings and machines, and what consequences that implies for things like machine rights and human-machine relationships.

In a 2007 essay titled "Religion for the Robots," I wrote that we would be likely to recognize robots as our equals if they developed religious inclinations, a claim I reprised in a full academic essay called "Religion among Robots" in 2024. The original essay was short, published in a newsletter called *Sightings* that was produced by the University of Chicago, and I was delighted to see in its aftermath that a number of Christian pastors found the essay so compelling that they used it as the basis of their sermons that week (some wrote about it online and some emailed me). Religious robots won't automatically be granted citizenship, but I think we're more likely to see robots as our equals if they've independently developed a sense of wonder at the universe and have begun asking fundamental questions about cosmic and personal meaning. I spent twenty years asking students whether they'd take a domestic robot to church/temple/whatever with them and saw those numbers shift from one or two students in a class of twenty-five to about half who would be willing to do so.

There are many connections between robot consciousness and religion: the matter is more than just whether or not robots want to attend religious services. The belief in future robot intelligence that grounds the AIproofing movement is, itself, rooted in several forms of religiosity. The futurists devoted to advancing AI technology claim their positions are empirically grounded, and thus, scientific; but there are many forms of faith at play. The first of these is rooted in AIproofers' belief in exponential logic and historical inevitability, both of which mirror Christian commitment to a divine plan. The fact that some technologies doubled in sophistication across some period of time in the past doesn't mean they always will or that such doubling would produce specific future outcomes. If past technological performance indicated future outcomes, then we would all – as the acclaimed innovator Bucky Fuller predicted we would in his book *Critical Path* – be teleporting from place to place. It certainly looked like transportation technologies improved exponentially when Fuller looked at them in the 1970s. But we cannot prove that our technological progress will be exponential or, even if it is, that there will be no end to it. We cannot prove that making machines fast will make them conscious or intelligent or human equivalent.

From its earliest days, artificial intelligence has been the locus of enthusiastic expectations among its researchers. The dreams of exponential growth are part of the persistent belief that human-equivalent machines are just around the corner. If we were to start "in the beginning," we would note how the founding myth of AI, the Dartmouth Summer Research Project on Artificial Intelligence, famously miscalculated the challenge:

> We propose that a 2 month, 10 man study of artificial intelligence be carried out during the summer of 1956 at Dartmouth College in Hanover, New Hampshire. The study is to proceed on the basis of the conjecture that every aspect of learning or any other feature of intelligence can in principle be so precisely described that a machine can be made to simulate it. An attempt will be made to find how to make machines use language, form abstractions and concepts, solve kinds of problems now reserved for humans, and improve themselves. We think that a significant advance can be made in one or more of these problems if a carefully selected group of scientists work on it together for a summer.

Among those who have tracked AI progress, there is widespread recognition that this founding document, written by John McCarthy, Marvin Minsky, Nathaniel Rochester, and Claude Shannon in 1955, is simply absurd in its underestimation of humanity and its overestimation of technological progress. To some extent, their mistakes promoted more caution in researchers as the decades unfolded, especially as the hyperbolic promises made from AI champions led to "AI winters" when poor results led to retrenchment in funding. But new models for AI hope resurface even in the darkest times, and the expectation of imminent machine consciousness was neither severed from nor particularly distant from AI research environments.

When I visited Carnegie Mellon University's (CMU) famous Robotics Institute in 2007, the language around artificial general intelligence (AGI) was just starting to percolate through AI culture. The term AGI didn't exist in pop culture yet, but AI researcher Ben

Goertzel actively promoted it within the discipline, and at least one member of CMU's faculty was getting emails from him about the launch of AGI conferences. At that time (and possibly still today), few people at the Robotics Institute believed that human-equivalent AI was plausible in the near-future even though Moravec had been their colleague prior to his retirement.

Moravec's former colleagues were, to my mind, charmed but bewildered by his religious pronouncements. When I told a gathering at CMU that I wanted to understand what led Moravec to write such a striking book, one declared "if you can figure that out, we'll all buy your book!" He did so with a wide and welcoming smile. It was not snide, but rather a recognition that something unusual, and often incomprehensible to other scientists, was happening. There were contemporaries, like Daniel Crevier, David Levy, Vernor Vinge, Danny Hillis, and most significantly Marvin Minsky (one of the "grandfathers of AI") who supported Moravec; but no prophet is welcome in his hometown. Or something like that.

People didn't object to Moravec's side projects, and I was explicitly told that *Mind Children* helped Moravec's promotion case (which surprised me). And there is no doubt that while I visited CMU Moravec was held in *far* more regard than the folks who echoed his claims; but still there was a fair bit of uncertainty around the entire issue. At the time, a graduate student said to me: "I'm glad that you note how most roboticists don't think about these things. Because we don't. I've never had a discussion about it with anybody." I followed up with him in 2025 (by which point he had established himself as a distinguished robotics researcher) and he noted of the Singularity and mind uploading that "academic roboticists don't really seem to take them seriously" and that his graduate students "really didn't engage with it all that much" when he included the material in a class on robotics and society.

Nevertheless, more people now find themselves persuaded by Goertzel, Moravec, and other futurists. The rapid progress in AI unleashed by OpenAI and other groups developing LLMs has changed the calculus for many people. Beyond the prevailing consensus in western nations, there are global movements adopting the language of AGI and the Singularity. A colleague in Africa has said to me "although not so many people are aware of AI, we are having proponents of AI singularity pushing for this narrative power-fully. It is a matter of time, as younger data scientists seem to be so obsessed with these ideas."

The LLMs do not actively think, and they don't really have conversations with people. Roughly speaking (and for the most part), what they do is predict the next most likely word(s) in a sequence based on a memory of what came before. They are an enormous leap forward in human-computer interaction; but they do not understand the world or the way their text fits within it. While a well-trained system provides coherent, often compelling text most of the time, it does so without understanding the relationship among the words. The LLM has a mathematical model for how words are connected, but not an understanding of the words or their connections. As such, no LLM will, by itself, become conscious in any way that stands up to real scrutiny.

Despite the factual impossibility of LLMs being conscious, there are many users who believe otherwise. In their 2024 study, UK scholars Clara Colombatto and Stephen Fleming surveyed 300 people and found that the more a person claimed to have used ChatGPT the more likely they were to believe it experienced things, and the more likely they are to believe that, the more likely they are to believe it could possess some degree of consciousness. More than anything else, this reveals the ease with which humanity ascribes agency to things in the environment. Like children whose teddy bears have thoughts and feelings, the users of a LLM believe it has experi-ences of the world.

Because interpretations of AI progress are faith based, the greatest concentration of people susceptible to believing in conscious LLMs lies at the intersection of those who are professionally involved in AI development and those who believe in the imminent arrival of AGI. In 2022, Ilya Sutskever, the co-founder and one-time chief scien-tist at OpenAI, declared on Twitter (the social media platform renamed X) that ChatGPT was "slightly conscious." It was an absurd claim to make, as ChatGPT was still nowhere near what neutral observers would call conscious. There is certainly some experience, as philosopher Thomas Nagel pointed out, which is what it's like to be a bat; but there is no such sense of what it is like to be an LLM. Its predictive model of text generation has very little in common with anything we know about human cognition and the many automatic errors built into its basic information processing ensures that only a committed believer, unconcerned by reality, could suggest otherwise.

And yet the hyperbole continued. In 2023, renowned AI researcher Geoffrey Hinton told journalist Cade Metz that "maybe what is going on in these systems is actually a lot better than what is going on in the brain." Not long after, Carl Franzen noted in *Venture-Beat* that when OpenAI launched GPT4o in 2024, the faithful were quick to label it "essentially AGI." Despite the enthusiasm, GPT4o was not, in fact, anywhere close to human equivalence; but, late in the year, OpenAI's Vaheed Kazemi announced on Twitter "we have already achieved AGI and it's even more clear with O1." It is not clear whether Kazemi believes GPT is conscious as part of his belief it is "better than humans at most tasks" (which, itself, isn't even remotely true). In mid-November of that year, after GPT scored well on the Abstraction and Reasoning Corpus, headlines rolled around like Geeky Gadget's "New MIT Research Proves AGI Was Achieved" (the article, itself, was fortunately more measured).

The repeated lessons in reality refuse to take root. Early in 2025, an anonymous AGI-advocate (@iruletheworldmo) wrote on Twit-ter/X: "grok 3 is here, and it's agi." Grok 3 was no closer to AGI than GPT, Claude, or any other LLM in 2025. At some point, we might build AGI, but the ever-present claims that it's right around the corner are nothing short of religious. There is little difference between them and two thousand years' worth of Christian expecta-tions that Jesus would descend from the clouds (including recent claims like those of inventor Joanna Ng, who told journalist Linda Kinstler that "Christ will rise before we see artificial super-intelli-gence"). These religious groups may be correct: either Jesus or AGI may emerge from a cloud; but either way the claims are built on faith, not reasoned necessity.

The utter credulity of AGI advocates bears striking resemblance to other victims of psychological manipulation. In a spectacular 2023 blog post, Baldur Bjarnason describes the parallel between people seeing intelligence in LLMs and an audience deceived by a public psychic. A psychic's audience is self-selected to be interested, prepared by the staging of the event (and likely observed for interesting data), narrowed down through the psychic's vague handwaving and statements, fooled by generic statements that sound "real," and then conned by sequentially more specific claims confidently asserted. Bjarnason describes this process and then notes that the exact same process operates among AI fans. Ultimately, he notes, it is even possible for many such fans to double down on the intelligence of the AIs because they see themselves as so intelligent that they *must* be correct in their estimation. All they've really done, however, is confuse statistically probable statements with deep insight.

This is not to say, however, that no AI will ever be conscious! Right now, I'm talking only the present day, not ultimate potential. Perhaps an AI will, indeed, become intelligent and conscious (which are not necessarily the same things). No LLM has attained consciousness of any sort by the time of this writing, and it seems impossibly far off until new technologies and approaches either replace or combine with the large language models like GPT.

For all that, anyone can see how conversational AI is getting much, much better by the launch of LLMs. The first chatbot was Joseph Weizenbaum's ELIZA, the most popular version of which purported to offer psychotherapy. Weizenbaum was very clear that he considered ELIZA useless as a therapist and was deeply concerned when psychologists suggested it or a similar technology might replace themselves. But there were users who could stay focused on a conversation with ELIZA for long periods of time. They would type something and then the program would parse the language, looking for a word that it can ask the user a follow-up question about. Occasionally, it would make declarative sentences, and these were liable to conversational confusion, as when it responded "you sound quite positive" when I input "yes, my dog died" as part of a brief and incoherent session.

Subsequent chatbots improved, with a variety winning the Loebner Prize, given annually to the best chatbot from 1990 to 2019. Steve Worswick's Mitsuku (renamed Kuki in 2020), which is not an LLM, won a record five times before the prize was suspended. None of these could convince a thoughtful human user that they were actually human beings, though progress certainly continued. Some years ago, AI researcher Ben Goertzel wrote: "at some point, some AGI research team is going to produce a computer program or robot that does something that makes the world wake up and say: 'Wow! Genuinely smart AI is not just a possibility; it's a dramatic reality! The time for humanity to create smart machines is now!' At that point, government and industry will put themselves fully behind the creation of advanced AGI." That time is, apparently, now.

ChatGPT quite took the world by storm, as the model was far superior to any previous chatbot. It made mistakes, including "hallucinations" when its predictive model guessed badly wrong; but it could swiftly, and generally accurately, answer questions and engage the user in a back-and-forth set of responses. Subsequent iterations of GPT improved on the model, providing more engaging responses and better information. In 2024, the GPT4o model had the ability to respond to spoken inquiries in addition to typed inquiries. It did not need to use a voice recognition software to convert spoken communication into text and then analyze it: it could directly analyze the verbal communication from human users. And as Goertzel predicted, a mighty rush of industry and government money followed.

However, neutral observers noted that the total amount of data, especially quality data, available to train subsequent generations of GPT was diminishing. There is only so much information-laden content on the Internet to absorb into the model (and there's lots of low-quality content in the mix, with more coming thanks to the proliferation of junk composed and uploaded by LLMs). Critics argued that human beings using GPT fill the Internet with GPT-generated content and that subsequent models would be "advancing" on the basis of the existing data plus dubious data generated by GPT itself. Meanwhile, even with the gains of techniques like retrieval-augmented generation, the models will never overcome their hallucinations or magically come to understand the world or their statements about it.

Despite these criticisms, faith in the chatbot revolution persists (though 2025's disappointing release of GPT5 may quiet the cheerleading). As noted above, such faith may be warranted over time. While it is unlikely that the specific technology underlying GPT can become conscious, it is possible that new techniques could supplement that technology and the resulting combination be considerably more than the sum of its parts.

Simultaneous to increasing processing speeds and chatbot capabilities, other computing technologies advanced. Whether this has been happening at accelerating rates is a matter of dispute; but there certainly have been improvements. So computers now compute faster. Unfortunately, if all these technologies do combine to form AGI, we may be in trouble. The "global race to AI supremacy" leaves little room for sensible maneuvering and we keep doing things to sabotage ourselves. For example, the departure of Miles Brundage, "Senior Advisor for AGI Readiness" from OpenAI in October 2024 is just one in a long line of moments that show we are improving our technology at a time when we diminish our investment in positive outcomes for it.

The AI narrative dominant in Silicon Valley culture takes past performance as evidence for future returns. According to that rhetoric, any trend apparent in the world of computers (such as Moore's Law) must be indicative of what will come in the years and decades to come. According to this narrative, then, it is inevitable that computers will soon equal humanity. And *then*, they will soon advance well beyond us. Imagine, for a moment, if Moravec, Kurzweil, and their followers are correct: computers double their ability every two years or so (remember that processing speed, memory cost, etc. are used as proxies for total computational ability on a scale that currently goes from zero to human-level computation). At some point a computer will equal a human being and then in just two years it will be twice "as smart" as that human being. Two more years and a computer will be four times as intelligent as a human being. And, supposedly, from there to infinity.

The cyborg response

It is necessary, of course, to believe that human thought is a form of computation akin to what computers do. Otherwise, the comparison doesn't hold. But let us grant this to our champions of the intelligence explosion. Without doing so, we cannot understand the fairytale we inhabit and we cannot be victorious (whatever that might mean). So, we shall presume that thought is a kind of computation and that the advances in digital computation will produce thought.

In this case, it will not be long before machines surpass human equivalence by a long margin. At that point, perhaps they will come to think human beings are a plague upon the Earth and decide we aren't worth having around. Or perhaps they'll simply take over all the most interesting opportunities, leaving humanity to wallow in the muck. No one can say for certain; but the mere possibility demands a response. What will humanity do if faced with irrelevance? What *can* we do in the face of such a relentless onslaught of computational superiority?

There can be no serious answer beyond leveraging the same technologies for our own gain. If the best thinking will be done by computers, then we will need to combine ourselves with computers. Arguably, this is already the case. The personal computer and the smartphone each, in their own way, extend the powers of the human mind in wondrous directions. A phone allows us access to information and to one another. It keeps track of memories and schedules, freeing us for (one would hope) more fruitful intellectual labor. A computer allows all these things and provides powerful tools for creating and sharing knowledge. These devices also provide opportunities for entertainment, play, and socialization. But our chief concern here is in the human potential to think, not apparent triviali-ties like our ability to enjoy life. So our computing technologies extend our ability to engage in what Moravec calls "meaningful computation."

If we want to accelerate that process, we will need to streamline the interface between our thought and the digital computation of our devices. Research into brain-computer interfaces (BCIs) has been underway since the mid-20th century, with a variety of academic projects and even consumer products designed to tap into human thought to activate a computer. For example, John Donoghue of Brown University works with patients who suffer from paralysis and Lou Gehrig's Disease. His company, BrainGate, creates BCIs that allow mental control over a robotic arm or restore speech.

As a leading figure in brain-computer interfaces, it is not surprising that Donoghue came to the attention of futurists seeking digital transcendence. Kevin Warwick, a UK roboticist, wanted to become what he called "the world's first cyborg" and he brought BrainGate on board for an experiment in which he inserted a computer chip into his arm and connected it to the Internet. Like Moravec, Warwick worried about existential risks before it was cool. But in Warwick's case, the machines Moravec loves are the source of his fear. In books and essays, Warwick echoes Moravec, suggesting that the speed at which computers and robots improve is fundamen-tally exponential and that the robots will therefore soon equal and then swiftly surpass humanity. He finds that worrisome rather than glorious.

In order to preserve human autonomy in the face of this "march of the machines," Warwick insists that we must become cyborgs. There is an obvious sense in which we already are: we outsource memory and communication to our smartphones, for example, and their loss cuts as keenly as though a part of us were removed. And drawing on cybernetics (the science of control in human beings and machines), anthropologist Gregory Bateson once suggested that even a far simpler human-machine system should be seen as one thing. Bateson argued that a blind man with a stick is one integrated system of communication "where no boundary line...can be relevant in a description" of the system. A blind man with a stick, then, amounts to what we call a cyborg.

Experimentally, Warwick put the cyborg hypothesis to an extended test. By implanting chips in his arm and using the internet to control a robot arm on another continent, he became Bateson's blind man with a 3000-mile stick. As he said to me: "once you're into the network of course this robot arm can be anywhere...what really shocked me when we did that was how that really changed things. The power that you've got. With technology, your brain and your body don't have to be in the same place anymore. If your body is technological, part technological, it can be wherever the network takes you. it can be on another planet. It can be anywhere. And that's part of your body. Your brain is controlling it directly and you're getting feedback from it." This new definition of the human self may become widespread if our species moves toward a cyborg future with built-in access to the Internet.

For Warwick, a technologically extended body and the potential for cyborg implants goes well beyond just controlling an arm: he wants something far more substantive. Warwick believes we must compete with AI on a level playing field, and that means a direct interface with silicon.

Human beings cannot think as fast or remember as well as computers. We cannot transmit information with the fidelity, reach, or speed of computers. But if we integrate our brains with computers, we could have their advantages. This raises obvious moral challenges, such as what to do in a world where some people cannot or choose to not join the cyborg revolution. Warwick argues we had best consider the moral and political questions of human-cyborg relations before they become reality.

In *I, Cyborg*, Warwick argues that "creating cyborgs completely changes the fabric of things;" it would be a "discontinuity, a nonlinearity, in evolution." Such a radical shift in the cosmos resembles the apocalyptic religious vision that the end of the world inaugurates a new one. Contrary to popular opinion, apocalypticism is not pessimistic: it is optimistic. Although often used to mean world-ending, religious apocalypticism is not built around the end of the world, it is built around the arrival of a new one. That optimism has skewed timelines for the traditionally religious (consider ancient Christians who believed the return of Jesus would happen in their lifetimes) and also for the 20th and 21st century futurists. In 2002, Warwick wrote, "with the interlinking of human and machine brains, even in a relatively limited way, it should be possible, within the next fifteen years, to upgrade memory, improve mathematical capabilities and increase considerably one's knowledge base."

In a 2012 conversation with me, I saw first-hand Warwick's enthusiasm for the work. While living in India, I was working on a project for a design magazine, which unfortunately went through some sort of internal crisis and closed up shop temporarily, killing my efforts before they could reach print. Reflecting on the future, and his uncertainty about it, he said "you're just going to jump off in midair and maybe you'll be able to fly and maybe we won't be able to fly...it's risky but that's a fantastic part of the experiment. I love it!" While progress hasn't met with Warwick's expectations, the decades following his experiments have produced many accomplishments. New efforts at brain-computer interface (BCI) technologies use implants and provide direct access to the brain (Warwick's experiments were in his arm, and in contact with the nervous system there). But whereas Warwick engaged his research from the angle of scientific and human exploration, recent efforts tend to be aligned with commercial interests.

Predictably, futureproofer Elon Musk stepped into the BCI business by purchasing a company with existing scientific and technological expertise. He bought Neuralink in 2017 and immediately accelerated its research timeline. Neuralink implanted its mesh BCI into the brains of monkeys, and then into their first human patient in 2023. The results for the human patient were initially remarkable, but they may have been temporary, as the majority of mesh wires retracted from their correct placement by mid-2024 (software improvements kept the system functional initially but public updates have been lacking). While the initial phase of Neuralink's research is therapeutic work on behalf of disabled individuals, the goal is to create a robust BCI that could be implanted in any human being. Such an interface would enable anything from mentally turning on the lights to driving a vehicle to accessing information on the Internet. In 2025, Musk promised that by 2029 the company would also implant its Telepathy mesh enabling mind-to-computer connection into some of the, supposedly, thousands of Neuralink patients each year.

Once a therapeutic implant of Neuralink captured the public's attention, Musk was quick to point toward the superhuman outcomes on the horizon for everyone else. For example, in an interview with Lex Fridman in 2024, Musk championed the rapid speed of brain-to-brain communication, which was one of the chief drivers of the new cyborg society predicted by Warwick. As Neuralink works to alleviate disabilities, it simultaneously opens the door to transhumanism. In Musk's jumbled together words: "a quadriplegic, or maybe have complete loss of the connection to the brain and body, a communication data rate that exceeds normal humans. While we're in there, why not? Let's give people superpowers."

Echoing Moravec, Warwick, and Kurzweil, Elon Musk sees human evolution heading toward a computational future. In his conversation with Fridman, Musk says that "the long-term aspiration of Neuralink is to improve the AI human symbiosis by increasing the bandwidth of the communication." In simpler terms, he wants to make it easier for human beings to connect with machines. Neuralink would remove the need for typing and for talking; it would (in theory) allow us to work with computers at the speed of thought.

While Musk's company received the lion's share of early 21st century press, other companies and other countries participate in the presumed integration of human and machine. Early in 2025, for example, China's state-owned Neu-Cyber announced plans to perform clinical trials with more than ten patients in the year ahead, with dozens of patients planned for the next year. For obvious reasons, research into overcoming physical disabilities will continue (and hopefully advance), and transhumanists seeking enhancement will follow that research, eager to participate as soon as they are permitted.

Cosmic redemption

Although some advocates limit their futureproofing to genetic engineering or cyborg prostheses, there is widespread Silicon Valley commitment to the belief that human beings will upload their minds into fully machine bodies. Based on books by Moravec and Kurzweil, this technofaith supposes that the new AI life will spread throughout the universe. The key will be a form of machine rapture: a metamorphosis that shifts biological humanity from its larval state to angelic immortality.

This is the future that I've elsewhere described as Apocalyptic AI. In the Apocalyptic AI mindset, our human limits in learning, memory, thinking, and lifespan could be overcome with digital technology. By uploading our minds into machines, we would become immortal and take on the abilities of computers. Eventually, we are supposed to spread computational intelligence across the universe. Modest versions of this exist, such as when Microsoft researchers Gordon Bell and Jim Gray suggested that we would eventually see "question-answering avatars" and these would "gradually become indistinguishable from the actual persons we know and love in 2001, enabling that person to 'live forever.'" But in the stronger, apocalyptic versions of AIproofing, the final cosmic transformation gets rid of the scare quotes, confirms the immortality of the individual, and revels in post-biological intelligence.

Kurzweil believes in six cosmic eras that constitute the entire history of the universe, past and present. Naturally, this version of cosmic history enshrines the inevitability of the AI apocalypse. He says that the first "epoch" was that of physics and chemistry, when nothing happened except things subject to the principles of those sciences. But thanks to the chemical combination of various molecules into complex systems, the epoch of biology followed. The first lifeforms evolved increasing complexity (which is really just an accident in Darwin's theory) until the point that brains evolved. This was third epoch. Using brains, animals developed technology, entering the fourth epoch. The most sophisticated tool user, of course, is humanity, and the most sophisticated technology that of the computer. By combining our brains with our technology, we will enter the fifth epoch. The acceleration of our thought and our power through the supposedly endless exponential doubling of computer intelligence will enable human/computer hybrids to spread throughout the universe. The sixth epoch brings computation to, as Hans Moravec refers to it, "boring old Earth" and beyond, until we have saturated the universe with posthuman machine thought. When that happens, Kurzweil predicts, the universe "wakes up."

Kurzweil's contribution to the Apocalyptic AI mindset has champions across the world. For example, in India his perspective has been taken up by scientists like VK Wadhawan and Govind Battacharjee. Both have published popular essays on our near-term path to becoming cyborgs and even uploaded minds. When I lived in Bangalore in 2012-13, I interviewed many scientists and engineers, finding that very few supported the Singularity or the idea that we could upload our minds into machines. But by the time I lived there again in 2018-19, those ideas were more common. And within a few years of my departure, there were academic and public events hosted in Bangalore to discuss them. For example, there was the 2020 "Facets of AI" conference at the National Institute of Advanced Studies, at which two separate speakers brought up the Singularity, and an event titled "Can Machines Come Alive?" hosted by Science Gallery Bengaluru in 2022.

The AIproofers understand human identity in a very specific way drawn from cybernetics and subsequent Apocalyptic AI thinking: they define an individual human being as a *pattern*. Each unique individual is a neurochemical information pattern, and the body is just the medium through which that pattern holds together. Usually, in fact, believers in the Singularity, mind uploading, etc. just think about our brains: they see our brains as running the pattern. In this, they avoid contemplating more complex possibilities, such as all the things happening elsewhere in our bodies, including in our bodies' symbiotic collaboration with billions of bacteria. We have bacteria on our skin and, especially, in our digestive systems that have direct impact on our moods and how we think. So, at a minimum, if human beings are patterns (not bodies, souls, or something else) then that pattern is properly constituted by something much bigger and much more complex than the brain alone.

Focusing on brains is sleight of hand, but the AIproofers might be correct in essence. We do not even remotely understand our brains, and we are even further from understanding how a human person might be the outcome of complex processes of brain, body, and bacterial colonies. But we may still learn enough about those things to mathematically describe the process. Of course, there may be something else to being human. Perhaps we have souls or perhaps we partake of some fundamental energy or universal consciousness, and perhaps machines cannot experience these. But that is neither here nor there for understanding the futureproofing religion. What matters is not whether they are correct or whether human beings are somehow more than physical description can provide. What matters is how the religion operates, both in the constitution of its beliefs and in the outcome of its work.

In terms of belief, pattern identity is a fundamental premise. The AIproofers see human beings as an information pattern and nothing else. If we are information patterns, then the body housing the pattern is irrelevant. Any body would do, including a machine body. Back in the 1950s, cyberneticist Norbert Wiener launched the pattern-identity position on the premise that all people are information, and that information can be transferred. He proposed in *The Human Use of Human Beings* that a person could, with sufficiently developed technology, be transferred via telegraph. From his theory of information to Captain Kirk's transmitter: beam me up, Scotty!

But the idea that we might reconstitute a person on the other end of a telegraph, telephone, or laser beam implies also the potential to restructure that person in new ways. So, transhumanists went a step further: they decided that if we wish to overcome the limits of biological bodies (and they did), it stands to reason that we should transition our patterns, our conscious selves, into computers. As far as I know, the first actual scientist to suggest that we could was the biologist George Martin, but it was the roboticist Hans Moravec who gave the idea its real intellectual weight.

Moravec's initial proposals suggested that we could make a rather abrupt transition, though he left room for subtle approaches also. In a 1979 essay for the science fiction magazine *Analog* and then in a watershed 1988 book, *Mind Children*, Moravec famously describes a robot surgeon connecting a human brain to a computer. As the surgeon cuts microscopically thin slices of the brain it fully copies the information processed by those slices and runs that information on the computer side of the connection. Eventually, it has sliced the brain up entirely, leaving a pile of mush, but the conscious thought of the person never ceased to function (Wiener had suggested, in very loose terms, something similar in his human telegraph transmission). The computer takes over each portion of the information pattern until it is ultimately running the entire show. On the one side, a pile of biological waste. On the other, a robotic angel.

The AI apocalypse brings a glorious new world in which we will attain glorious new bodies. Where ancient Jews and Christians foresaw a new world created by their god, this computational version emerges from technology. In *Robot: Mere Machine to Transcendent Mind*, Moravec says that "technical civilization, and the human minds that support it, are the first feeble stirrings of a radically new form of existence, one as different from life as life is from simple chemistry. Call the new arrangement Mind. Unlike Life alone, which learns from its past but is blind to its future, Mind can choose among alternatives to imperfectly select its own destiny – even to amplify that very ability." The new world is one where we take up a new mode of living, no longer subject to the whims of chance, or even necessity.

According to the logic of futureproofing, this new form of life protects us from extinction. Well before existential risk became a rallying cry, Moravec wrote of intelligent machines that "like biological children of previous traditions, they will embody humanity's best chance for a long-term future." All that we require is to give up our biological commitments and enter the machine age.

Of course, most people don't agree with Moravec that their biology is "mere jelly." Most folks wouldn't get too excited about a surgery where the robot surgeon leaves behind a corpse. So Moravec noted that we could also use high resolution brain scans or lifestyle recording devices to capture a person's identity. Almost all of the subsequent AI proofers preferred the idea of high-resolution scans.

Ray Kurzweil is one of the advocates who thinks that we can develop brain scanning technology to such resolution that we will be able to identify the neurochemical process at stake. He believes that as we advance our cyborg prostheses, we will gradually upload ourselves to digital life by shifting more and more resources to the nonbiological part of our intelligence. His argument is much like the classical ship of Theseus argument: is Theseus in the same ship or a different ship if he has replaced every piece of the ship one by one? That is, we know it's a new ship if we take a pile of lumber and hard- ware, build a new ship, and have Theseus walk a gangplank from his old ship to his new. But what if we pull one board out of the old ship and replace it with a new board? It's still his original ship, right? Well, in Kurzweil's position, if we do that over and over, replacing each board, mast, sail, and rope with a new one then, at the end, we still have Theseus's original ship. After all, it might have taken us ten years before each one of those parts was fully replaced. No one would suggest that at some point along the line, the ship no longer belongs to Theseus. This logic certainly applied to rebuilding my backyard deck when I lived in New York: no permit required if I were to replace it board by board, but to tear it down and rebuild it begins with bureaucracy. Similarly, Kurzweil believes that if we replace biological parts one at a time, then we never stop being ourselves.

And, says Kurzweil, if that's true then it implies undergoing brain and body replacement all at once *also* leaves us with our original selves. Of course, that means there would likely be both a biological person and the postbiological person, each of whom claims ownership over the car and commitment to their spouse(s) and children. It's not without reason that Kurzweil says nonbiological consciousness will be a major legal and moral concern!

On top of the problems a person would have after choosing the Moravec operation – whether a surgical operation or a high-resolution scanning of brain patterns – other problems emerge from the fundamental idea that replicating a pattern creates that person in a literal way. One primary goal shared by Moravec, Kurzweil, and others is the accumulation of enough data about the deceased to "resurrect" them in machine bodies. The resurrection of the dead is another of the obviously religious aspects of the futureproofing agenda, and it drives real people's behavior. Digital resurrection certainly raises legitimate legal, moral, and social problems, much of which I reflect on in my first book, *Apocalyptic AI*.

We've already seen Nikolai Fedorov, a Russian thinker who felt it was the Christian obligation of humanity to "resurrect the Fathers," going all the way back to the Biblical Adam. As Fedorov argued on behalf of this task in the early 1900s, there is quite a long pedigree for the idea that we can and should use technology to bring back the dead, potentially all of the dead. Fedorov requires that we collect a person's atomic remains, but Moravec isn't otherwise far from Fedorov's vision. Moravec supposed that we could use computers to generate infinitely high-fidelity replications of any period in the past, thus resurrecting our ancestors, and perhaps running through history over again or changing it to see what happens in the "simulation."

Among his other ideas, Moravec is also the first to publish on the "simulation hypothesis," the idea that it is statistically more likely we are living in a simulated universe than in the physical, base reality. Until recently, credit for this has been frequently given to a philosopher in the UK who claimed priority even knowing that the idea is not his; but fortunately, more academics and journalists have become aware of the idea's pedigree. Moravec, however, may not have suggested the simulation hypothesis first either, or, at least, not by himself. Engineer and space advocate Keith Hensen noted in a 2006 email to Eliezer Yudkowsky's SL4 listserv that in 1986 he joked with Moravec about having previously had their current conversation many times already because any society that could produce massive computation might replay history in simulation. Assuming this story is true (and I have no reason to dispute it), it implies the entire panoply of very committed and serious conversation around the simulation hypothesis (so serious that Neil deGrasse Tyson hosted a wildly popular event about it at New York City's Rose Planetarium) – this serious conversation began with a joke. Then Moravec articulated it and the possibility it raises for things like resurrection, and from there it reached Silicon Valley.

Thanks to advancing chatbot technologies, we can legitimately glimpse the possibility of something like computational resurrection. In July of 2024, Max Zahn reported on the booming industry of "grief tech," in which users upload data from text messages, letters, etc. to the training set for a chatbot. Such efforts draw directly on the decades of work on "mind files" conducted by sociologist William Sims Bainbridge, especially his collaboration with the group Terasem. When Bainbridge and Terasem began working toward the resurrection of mind files, it was through a very long questionnaire, and it was with the goal that someday a computer could use the data. But they weren't the only people invested in chatbot resurrection. In 2016, journalist Casey Newton described the first real effort at resurrecting someone using a chatbot. The technology had limits but even then could provide the sense that a deceased individual has "taken a new form."

But such technologies are moving fast: early in 2025, researchers from Stanford and Google published their efforts revealing (by their measure) an 85% accuracy for digital replicas after users spent two hours interviewing with them. Should innovators combine LLMs with other technologies to eliminate chatbot hallucinations there may, in fact, be little difference in conversation with the original human being and conversation with their digital clone. As Bainbridge once said of his character in the game *World of Warcraft*: "I would consider a continued existence for my main WoW character, behaving as I would behave if I still lived, as a realistic form of immortality...Ultimately, virtual worlds may evolve into the first real afterlife, not merely critiquing religion but replacing it."

Ray Kurzweil, one of the staunchest advocates for a post-Singularity future, openly acknowledges his desire to resurrect his father using digital simulation. In the documentary *Transcendent Man*, he reveals a veritable trove of records, personal letters, and effects that he stores until such time as a computer with sufficient computation can take all the artifacts and render a reasonable "imitation" of his father. Kurzweil claims that imitation will, if fully accurate, *be* his father. This reveals an important aspect of pattern identity, throwing into stark relief how "pattern identity" differs from things like body identity (the idea that I, as an individual, am this flesh and bone person). In recent work, Bill Bainbridge has attempted exactly such a feat, engaging with chatbots he trained on his ancestors' writings. Meanwhile, sites like hereafter.ai want to bring simulated resurrection into everyday life.

A demonstrably identical digital resurrection or mind upload conjures substantial legal questions. Does the resurrected individual have any claim on previous property? What rights has it re-acquired? Does it get to vote? Make medical decisions for a living spouse? Retract end-of-life decisions made by the biological self? Does it owe money to creditors? These problems and more will emerge if a digital vision of resurrection and immortality comes to pass. One rather strange legal maneuver is already underway: in 2020 the US Patent Office issued a patent to Microsoft for "creating a conversational chat bot of a specific person" (patent number 10,853,717B2). How that could play out in a world where Microsoft did not invent the idea and many groups are attempting the feat remains to be seen. The capacity of LLMs to copy a person through predictive text probably circumvents the Microsoft patent, but perhaps the limits of LLMs lead back to the patented architecture...and corporate control over the deceased.

Problems, however, are simply stumbling blocks to be overcome in pursuit of the higher calling. Just as Fedorov referred to the resurrection of the dead as the Common Task, AIproofers argue that we all have something to gain by transferring our consciousness into machines: individual immortality and the preservation of the species. If we can upload our consciousness into machines – whatever form that operation takes – then we can utilize our machine bodies to escape Earth, and to avoid any and all threats that would eliminate life here.

Glory be

As supposed in many religions, there is no stopping the salvation promised by AIproofers. The law of accelerating returns, extrapolations of Moore's Law, assumptions that evolution guides thought from biology to digital computation are all variations on a theme: that the future of the universe is digital and there is nothing to stop this progress toward the fulfillment of cosmic purpose.

Ultimately, the pursuit of immortality is supposed to produce the kind of transcendent gods that traditional religions describe in support of their own claims to salvation. Kurzweil writes that "evolution moves inexorably toward our conception of God, albeit never reaching this ideal." He believes that "once we saturate the matter and energy in the universe with intelligence, it will 'wake up,' be conscious, and sublimely intelligent. That's about as close to God as I can imagine." Motivated in part by fear that human beings will go extinct, Kurzweil proposes we use AI to generate human salvation, cosmic purpose, and the birth of the gods.

In my scholarly work, I have repeatedly noted the optimistic worldview of apocalypticism, and also how this impacts the time scales of AIproofers. For Kurzweil, "most of the readers of this book are likely to be around to experience the Singularity." His prediction closely resembles the Gospel of Mark, which promises "they will see 'the Son of Man coming in clouds' with great power and glory...and gather the elect from the four winds" before further proclaiming that "this generation will not pass away until all these things have taken place." Naturally, Kurzweil is not far from his predecessor, Moravec, who has said computer intelligence will equal humanity by 2030 and significantly surpass us by 2040. From here, intelligent life will reject all limitations, pursuing some form of absolute transcendence. Moravec predicts that once we have uploaded our minds into machines and left the earth, "a 'Mind Fire' will burn across the universe. Inside the Mind...physical law loses its primacy to purposes, goals, interpretations, and God knows what else."

These promises hold powerful sway in Silicon Valley. Just glancing at a couple of photos of OpenAI's library, published by *New York Times* journalist Cade Metz, reveal the presence of A.C. Clarke's *2001: A Space Odyssey* (which influenced Moravec and practically everyone else) and Moravec's own *Mind Children*. Without visiting myself, I'd take a bet at almost any odds that Kurzweil's *The Singularity is Near* also graces the shelves, and possibly another of his books. It would be interesting to know how many more pop science and SF predictions of robot immortality are present. Alas, the photos do not reveal copies of my own books. Surely, at least *Apocalyptic AI* deserves a spot!

Whether we head toward an angelic future remains open to debate. The advocates of intelligence explosions, the Singularity, and mind uploading speak in definitive statements about inevitable futures. But for all that, the future has not been decided. Human beings still have decisions to make. Certainly, we should act swiftly – much swifter than current governments, religious groups, and citizens have chosen – to corral AI dangers and protect humanity. Whether we should apply equal vigor to the pursuit of superintelligent AI and uploaded human salvation is less clear.

For AI advocates, however, the proposition that we could become transcendent machines carries great weight. It offers redemption from the present, proofs the future against our demise, and delivers the peace of mind that our world has a purpose, one that includes a glorious new life for our species. All that remains is to consider our options for reaching that point and beyond: how will humanity live as it pursues these transcendent dreams on Earth and in space?

CHAPTER FOUR
GHOST TOWNS

The remains of human habitation dot the landscape, but these do not occupy our dreams so thoroughly as the homes we have not yet built. Passing through mountains we think of old mining towns, long since abandoned when the steel mills closed or the rivers ceased flowing with gold. Reaching into the past, we think about Cahokia, of Chaco Canyon, of Pompeii, of Troy, of Mohenjo-Daro, of Angkor, and of Tanis (especially if we are fans of *Indiana Jones*). The list goes on. The ruins of the past remind us that no place is forever, that people die and societies collapse. Like the American Old West, history is littered with the skeletons of cities and towns. There's a lingering charm when we learn about them in school; but with fierce vigor we dream of new homes, new cities. If fallen cities assert our mortality, then the futureproofing religion – the commitment to a forever people – must rethink the landscapes where people (or machines) live.

In part, we must imagine new cities because our current cities fail us. Our headlong rush into climate change, into automation, into reconfigured biologies, is a direct outcome of the way we choose to live. If we worry that humanity is at risk, then we worry about the sturdiness of our shelters. This is not a new dynamic, though its scope may be somewhat more cosmic than in the past. The grandeur of distant cities lies at the heart of Plato's Atlantis and Marco Polo's meeting with Kublai Khan. In his 20th century masterpiece, *Invisible Cities*, Italo Calvino resurrects those glories, retracing the path of Marco Polo and offering a new vision of cities by revealing the things that make a city but that we cannot see: the relationships that transcend the buildings. The weight of time, the precarity of existence, the exchange of goods and family ties, the governance and the gossip. Through Marco Polo and Kublai Khan, Calvino finds the axis where our imagination of a city and the literal-ness of the city become one and make the city real.

It may seem strange to anticipate the future by looking to the past, but the visions of what a city could be reveal the heart of futureproofing through urban design. A city of immortals must reckon with the mortality of cities. The relations of angels, even digital ones, may not appear except through buildings and their byways.

Perhaps we have always sought new living arrangements but lacked the means to share our visions widely. Copies of *The Travels of Marco Polo* were limited by the speed of the printing press and the wagons or ships that carried volumes from person to person. Recent pioneers have longer reach. Stewart Brand built an entire movement around his *Whole Earth Catalog*, a movement that changed the way Americans – and future Silicon Valley entrepreneurs – viewed their world and their opportunities within it. Famous for his "access to tools" philosophy, Brand stands at the fulcrum of our old and new ways of living. As chronicled in John Markoff's *Whole Earth*, Brand drew on indigenous American traditions, emergent communes, the counterculture, and the rise of digital computing as part of his own utopian vision. Is it any wonder that he joined the 21st century project to resurrect mammoths in Pleistocene Park? Ultimately, Brand gave impetus to both back-to-land communes (not that he wanted to live in one) and communities of freethinking, tech positive urbanites.

Some of us – myself included – yearn for a world where we can live surrounded by nature, perhaps even living off the land; but the future of humanity is in cities. The communes of the 1960s cannot hold us all. Most of them didn't even survive the 70s. There are simply too many people to spread out as far and wide as we may choose, and there are too many desired conveniences for us to live in isolation. Urban and suburban density become the solutions to the growing populations across the world. But our roads follow the paths trodden by 19th century cattle and our infrastructure looks more like ancient Rome than the planet-wide *Star Wars* city of Coruscant. So our future cities are somehow yet to be realized.

Cities in science fiction. Cities in pop science. Cities in transhumanism. There's an overlapping Venn diagram, and urbanproofing sits in the central position. Near the end of his life, Julian Huxley declared that "the outstanding social habitat is the city, the habitat of civilized man." But recognizing that urban design all too often fails us, he also speaks of attending to the "social ecologies of cities." He knew that in the transhumanist world of personal and collective fulfillment, urban design would matter. This is particularly true as we increasingly rely on cities for modern life, simultaneously lured by their attractions and driven by the collapse of opportunities beyond their borders.

Rapid changes to the Earth's environment will necessitate increasing density as people look for efficiencies in resource management and perhaps witness fewer places suitable for human habitation. In Bangladesh, for example, up to 30% of available farmland could be underwater by the middle of the twenty-first century. With a majority of the country's land below sea level, and thus at risk of being permanently submerged, rising ocean levels are of direct concern. Obviously, Bangladesh is not unique in witnessing the threat of rising sea levels or other consequences of climate change.

Changing weather patterns, rising sea levels, frequency of natural disasters: as bad as they are, the greatest impetus for urbanization is really economic. People flock to cities for opportunities. Real estate prices push people away, but jobs, cultural opportunities, shopping, and lifestyle cachet pull them in.

This conflict explains the 21st century movement toward relaxed zoning laws. In New York City, for example, the Basement Apartment Conversion Pilot Program provides financial assistance in converting unused basements into safe and legal apartments. Similarly, laws to loosen zoning restrictions and allow additional housing like backyard cottages or garage apartments – could provide new housing opportunities without requiring large housing developments.

But there are always squabbles: increasing suburban density puts pressure on schools and roads. Building affordable housing terrifies the affluent. And as the cost of urban housing reaches toward impossibility, low-income individuals and families face the time tax of long bus commutes or, far worse, homelessness. And yet too few job opportunities exist in the exurban towns and, especially, the hinterlands beyond. In the midst of such economic chaos, cities struggle with waste management, infrastructure, even rat infestations. For the rich and the powerful, the financial and technological elite whose own future is perhaps the only one they actually wish to preserve, American cities too closely resemble those of developing nations. For these reasons and more, the religiously technological believe that cities must be rethought and new urban paradises built.

There is no sharp line of distinction between the design of high-tech future cities and the design of high-tech future lifeforms. The futureproofing agenda dreams of reconstructing human bodies, human minds, and human habitations. In this chapter, we look at how the elite tech community, bolstered by the enormous financial returns of an ever-widening economic gap, promise a new way of living for all humanity.

Cities without limit

The entrepreneurial class has joined forces with a variety of political and commercial actors to reinvent urban design. Rather than look toward improving local facilities, they cast their eyes wider, toward "open frontiers" that can be urbanized. These urbanproofers probably will not have homes for everyone, though their language often purports to universality, because they propose new cities in the deserts and the high seas, futuristic utopias for tech aficionados.

For whatever reason, reimagining political affiliation follows the reimagining of cities. For many of the same tech entrepreneurs eager to reconstitute urban experiences, there is a connected interest in drawing up new, entrepreneurial nation-states. It's probably no surprise that one of the leading "we need a new nation" folks is also a leader in the cryptocurrency marketplace: Balaji Srinivasan. As Gabriel Gatehouse notes for the BBC, in Balaji's universe, "you would choose your nationality like you choose your broadband provider. You would become a citizen of the franchised cyber statelet of your choice."

This notion of a franchised government draws on the same cyberpunk science fiction narratives that created so much of the contemporary tech ethos. In *Snow Crash*, Neal Stephenson describes a Greater Los Angeles constituted out of so many little independent nations, from the Mafia and its famed Cosa Nostra Pizza franchise ("The Mafia: you've got a friend in the family!") to Little Hong Kong. Even the United States of America still has a small enclave. In Stephenson's future, being a citizen of one of these communities grants entry and privileges. Following this, the role-playing game *Shadowrun* provides no doubt that corporations have become nations unto themselves, with militaries, laws, and, in some cases, sovereign territory. And in a moment of fiction-turned-reality, the British Secretary of State for Science, Innovation and Technology, Peter Kyle, said in 2024:

> I'm probably the first secretary of state that is dealing with companies which are outspending our entire British state when it comes to investment in innovation. So let's just act with a bit of sense of humility. We are having to apply a sense of statecraft to working with companies that we've in the past reserved for dealing with other states.

For Kyle, that seems to mean backing away from regulation and trying to ensure that tech companies don't move their products from unfriendly regulatory environments. But it hits close to home when tech entrepreneurs start discussing their own nations.

In his book, Balaji (according to Gatehouse he prefers to be known by just that name), says it is possible to create a conceptual nation that, in time, finds its way onto *terra firma*. Amidst a long litany of complaints about the *New York Times*, what he perceives as "woke ideology," and an imagined group of American leftists who supposedly adore the Soviet Union, he hopes to reconfigure the world order around Bitcoin (surely without regard for his own profit margin if that happened). "When we think of a nation state," he writes, "we immediately think of the lands, but when we think of a network state, we should instantly think of the minds" and so "a key concept is to go cloud first, land last." Starting from a rather hilarious commitment to "the absolute truth of the Bitcoin blockchain," Balaji believes that people can unite around a single idea (he probably pulled this bit of foolery from Ayn Rand, and his suggestions for potential ideas are somewhat scattershot) and then have bitcoin and a social network to unite them with such vigor that they'll buy some land, creating a "network archipelago." The community would somehow force the actual governing nations, nations providing the people with basic services like roads and electricity, into ceding control to this new group of the wealthy (median income in network states is supposed to be $200,000 per year in 2023 dollars, compared to that year's U.S. median income of $65,000 and a global median that would be perhaps one tenth of that).

Balaji wants a pre-existing social network to buy land and become a government; but most urbanproofers want to move straight to the land and recruit citizens. One such "nation," Próspera, is an island off the coast of Honduras, and has quasi-permission to write its own laws. By the turn of 2024, the Honduran government began carving into Próspera's power, but the outcomes of that remain to be seen. Off the coast of Honduras or elsewhere, wealthy investors in urbanproofing seek immunity from the laws that govern mere mortals. They wish for new societies unburdened by the laws and taxes that ordinary citizens accept as necessary parts of life. In such a city, urbanproofers suggest, they can advance technology and ameliorate some of the crises of modernity.

While many of the urbanproofers who wish to redesign the city experience trace their ideology to the libertarian urban planners of the mid-20th century, they exist in a longer lineage. Religious reformers often head to the wilderness, where they can build communities in line with their values and vision of the future. Of course this includes Kool-Aid drinking cultists, but it also includes the Oneidans and the Shakers. In American history, these groups founded communities that aspired toward a better world, often establishing new laws and frequently envisioning new industrial practices (sometimes for religious reasons and sometimes more hedonistically). The urbanproofers obfuscate their religion in a fog of technological necessity; but they are just as much committed to a world of faith, ritual, and superhuman realities as their traditionally religious forebears.

Technofuturism

Governments and private capital have entered a new phase in urban design: independent futuristic cities. Chasing everything from smart city status to tech investment, local governments have endorsed what Hannah Rebentisch and her coauthors label unicorn planning: "a techno-optimist approach to urban and economic development that prioritizes the creation of instant 'smart' cities." These cities promise a reorientation toward the natural and human environment while calculating their merits on the basis of high risk and potentially high return value. One 2nd century Christian theologian has been famously misquoted as saying "I believe because it is absurd." I don't know whether that is relevant to traditional religion, but it definitely applies to high-risk entrepreneurial urban design. Faith that urban life can be reconfigured as a brand new techno-utopia draws on hope for sustainable energy structures, superior human life balance (e.g., the "fifteen-minute city"), and the integration of natural beauty accessible to all city residents. Simultaneously, these new designs almost universally posit new legal structures, rejecting the previous forms of government in favor of either libertarian self-rule or some other complex of economic/legal jurisdictions that promises salvation from the malaise of modern society.

Saudia Arabia's NEOM is among the headline projects of the 21st century though every year seems to scale back the rhetoric. Drawing on "neo" for new, the city's webpage further points to the name meaning "new future" as the letter m is the first letter in the Arabic word for future, *mustaqbal,* and the fact that m is also the first letter in the name of Crown Prince Mohammed bin Salman. In 2017, the Crown Prince described the city-to-come as "drone-friendly and a center for the development of robotics...a place for dreamers who want to create something new in the world, something extraordinary." Marc Raibert, CEO of Boston Robotics, elaborated by predicting that NEOM's robots would do everything from security to eldercare. NEOM's own birth comes through a robotic midwife: late in 2024, the Saudis launched an engineering initiative using cutting-edge construction robotics to build rebar cages for concrete. The bones of the city knitted together by robotic "hands."

In advance of NEOM, Saudi Arabia had already committed to robotic and AI technologies. In 2017, for example, the nation "granted" citizenship to the robot Sophia, a product of Hansen Robotics. The announcement, presumably more public relations than social re-engineering, is part of the nation's strategic position for 21st century technology and culture. Simultaneous to announcing a society of robot servants in NEOM, Saudi Arabia also gave a passport to Sophia. Clearly, Saudi Arabia's leadership sees the future in terms of urban redesign and the inauguration of a paradisical techno-wonderland but isn't overly concerned with ideological consistency.

Naturally, the city-state has its own website to encourage invest-ment and interest, and which describes NEOM as a "revolution in civilization" and as "the land of the future, where the greatest minds and best talents are empowered to embody pioneering ideas and exceed boundaries in a world inspired by imagination." The utopian vision of NEOM hews closely to the economic propaganda of Silicon Valley entrepreneurs. Built by robots, inhabited by robots, and ready to incubate the next generation of technological futurism.

But NEOM is also about humanity and our shared future. As one of its promotional videos asserts, NEOM aspires to be "a land for free and stress-free people; a start-up the size of a country: a blank sheet of paper on which to write the new era of human progress." This kind of freedom echoes centuries of utopian hope, now worded in the entrepreneurial language of Silicon Valley.

Primarily within the realm of religion, reformers have proposed that new habitations could create new versions of human life. This is not necessarily wrong. After all, the famed architect Le Corbusier defined a house as a "machine for living in," and the same would be true for a city. He decried the fact that we have "not yet thought of building houses adapted to ourselves," and the corollary is that perhaps we have not yet built cities suited to our present, and certainly not to our future. The cities we have, from such a vantage, constrain our lives and our vision. Religious reformers up to and including futureproofers certainly commit to this position. As we all know, machines are tools for meeting our goals, but they simultaneously shape the way we see the world. As in the adage, to one with a hammer every problem looks like a nail, what we have at our disposal both limits and reveals what we see in the world. For the urbanproofers behind NEOM, a new city, a new machine for living in, means a new kind of human world.

NEOM might find a sister city in Telosa, the vision of billionaire Marc Lore. According to its website, Telosa will create "a more equitable and sustainable future." Telosa's governing board describes it as a "new model for society," where "everyone has an opportunity for unlimited growth." Like NEOM, the plans for Telosa would enable better access to nature and walking convenience of all necessary businesses and education; but the website adds a focus on safety and inclusion for all people.

On their webpage, Telosa reports that they are "absolutely not attempting to create a utopia" because "utopian projects are focused on creating a perfect, idealistic state." Here they echo the transhumanists who speak of endless progress, specifically because there is no final salvation in transhumanism (or anywhere in the futureproofing community). And yet every transhumanist group – and now the futureproofers also – must contend with the fact that everyone else sees what they're doing as utopian. If it acts like a utopian fantasy and talks like a utopian fantasy...

Telosa's denial of utopian aspirations garners little traction because the organization directly reconfigures its economy around a rather utopian economic system: "equitism." This philosophy revolves around land ownership by the foundation and supposedly provides better economic opportunities for all (in keeping with this, Telosa has a VP of diversity initiatives who aspires to ensure the community benefits are shared beyond white Silicon Valley types – it's an open question whether this job will survive in Donald Trump's America). Rather than private ownership that benefits the few, foundation ownership supposedly reinvests property value growth into the systems all citizens rely upon.

Given that one cannot sell the land, there's no obvious sense for how its increased value would benefit anyone or what it would even mean for the land to increase in value. One is permitted to buy and sell homes...which is a proxy for land value, but does it mean that profit goes to the foundation? The Telosa effort to rebuild capitalism seems dubious, and one might similarly wonder about the feasibility of cities establishing their own laws. The economic and political agenda is, therefore, fundamentally one of faith. Whatever the founders may believe, they expect the adherents to invest themselves in a utopian social structure that promises an honest and productive environment for all residents.

At some point, one almost feels relief at the honesty of Silicon Valley entrepreneurs who want to build a city in northern California because, in their own words, the "financial gains could be huge." California Forever began as Flannery Associates (backed by Marc Andreesen Horowitz, Reid Hoffman, Laureen Powell Jobs, and other business leaders and venture capital outfits); the organization bought land in Solano County, raising the ire of environmentalists, farmers, and other locals. They sought to reverse this with an implication, not a promise, that the organization might provide down payment assistance for locals who purchase homes in the new city and further that the organization will protect local interests. But their promises fell on deaf ears. In 2024, the organization withdrew a ballot initiative on its planned city and alleges it will return to the public after environmental review and better outreach/engagement with locals.

Little can be said of California Forever beyond recognizing a litany of powerful backers and a local populace in the early stages of revolt. As with similar projects, the city founders allege they will build a walkable, environmentally friendly urban space with economic opportunities.

Some future cities would obviate the need to fight over land because their cheerleaders dream of living on the ocean. This "seasteading" movement garnered attention thanks to backing by billionaire Peter Thiel. While aircraft carriers and cruise ships are already cities on the open water, the seasteading movement desires something more: permanent, self-governing habitations. There are literary and pop science antecedents, but it was the collaboration of Wayne Gramlich and Patri Friedman that brought tech money to the game. They launched the Seasteading Institute, which advocates for further development.

At present, no meaningful seastead can be imagined outside of territorial waters, and so the early efforts associated with the Seasteading Institute have all been under the auspices of government arrangements. Alas for the movement's advocates, however, (vaguely) supportive governments in French Polynesia and Thailand have reversed course and become antagonistic to the seasteaders. Given that seasteading ultimately rejects terrestrial governments, the conflict seems more or less predestined. In a 2009 article published by *Wired,* Chris Baker quotes Patri Friedman's disdain for current government and the impetus to build new cities: "government is an inefficient industry because it has an insane barrier to entry. To compete with governments on existing land, you have to win a war, an election, or a revolution." Balaji later quoted this in advocating for network states.

And thus Friedman dreams of libertarian experiments on oceanic cities. Even if the technology proves intractable, the advocates of seasteading (and its post-territorial, antigovernment stance) have opportunities to produce relevant policy. In 2025, Jim O'Neill took over the chaos-infused Center for Disease Control in the U.S. after previously working with Peter Thiel on his seasteading initiative. Whether the mix of new habitats, new (or no) governments, and biotechnological immortalism that reigns in urbanproofing will gain strength through O'Neill's position is unclear; but there is at least an opportunity for both the ideology and the startup companies involved (O'Neill also helped Thiel with an investment fund).

Building a city on the ocean will require even more technological advancement than what urbanproofers dream in the NEOM and Telosa projects. But tech development is, as we have seen, crucial to the entire futureproofing agenda. Naturally, the visionary work for these ocean habitations leverages public interest in futuristic environments. Artwork tends toward geodesic domes sitting on platforms with intriguing geometric configurations or the sort of modular building construction that looks like a cross between postmodern architecture and a Jenga game about to collapse. Living platforms rise above (surprisingly calm) seas and luxury yachts anchor nearby.

Given that the futuristic cities – on land or sea – are wrapped in dreams of untaxed, visa-free tech incubation, it should not surprise anyone that they connect directly to the broader promises of futureproofing. We have already noted NEOM's investment in robotics and the broad interest in energy security that permeates urbanproofing. An even stronger example would be Próspera, which aims to be a hub for a regenerative medicine clinic. Peter Thiel and Sam Altman are both advocates of and investors in Próspera, and the city aligns with their technological and personal goals: it promotes fantasies of eternal youth and technological paradise. In late 2024, Próspera's website stated:

> Próspera is seeking proposals from qualified practitioners and companies to establish a state-of-the-art clinic for medical therapies, clinical trials, and treatments, with a special focus on regenerative

medicine and longevity therapies. Próspera offers a jurisdiction opti-
mized for medical innovation, providing an unparalleled environment for pioneering medical ventures. The primary objective of this
RFP [request for proposals] is to invite proposals to create a clinic
that will provide clinical space for companies conducting clinical
trials and delivering therapies.

The urban landscape is thus re-planned in a comprehensive effort to
prevent death. In this way, it cannot be divorced from the broader
futureproofing movement.

The dream of (near)immortality has long accompanied the
building of new cities, even if it looks unfamiliar to recent history. In
The New Atlantis (1626), Francis Bacon described an imaginary island
of Bensalem, where Christians used technology to fulfill the divine
plan for the world. Theology constrained Bacon: he could not simply
disregard Christian promises of heavenly redemption. But in spite of
this, he saw technology as divine and supposed that a properly
constructed community would use technology to improve energy
access and agriculture while enhancing human health and also
longevity. Not so far from Bensalem to Próspera!

Desert blooms

Strangely, many of the techno-urban dreams are imagined in
desert landscapes. Cities like Phoenix struggle with water access,
draining rivers and endangering wildlife and smaller human communities. Despite this, the urbanproofing community believes they can
create environmentally sustainable cities in places with little to no
water available. Recognizing that car culture, traffic, and pollution
characterize the early 21st century urban experience, they re-envision
cities as places where people live in harmony with nature, helping
counteract climate change, in cities often founded in environmentally counterintuitive locations.

The dream of desert cities exerts a siren call for many nations.
Scholars Laurence Côté-Roy and Sarah Moser describe the new cities

underway in Morocco, the outcome of tens of billions of dollars, designed to handle population, spur economic growth, and advance Morocco's investment in digital technologies, green energy, and sustainable agriculture. So, it's not just the family of tech companies and billionaires who will pay for urban redesign; but, in some cases, public funds are a part of that mix. In such cases, the public coffer never opens in isolation; instead, there are collaborations with private industry that reflect an uneasy consensus as to how humanity must prepare for the future while solving a variety of present-day problems. In Morocco, several companies operate in a no-man's-land of absent jurisdiction and regulation, where it is hard to evaluate the extent to which these corporations adhere to government and public objectives. That problem perhaps explains why so many of the urbanproofing projects are planned for remote landscapes. No one is there to see what happens.

From Arizona to Saudia Arabia, desert landscapes often come cheaply; or, at least, cheaper than the alternatives. Home prices in Phoenix are on the rise just like everywhere else in the United States; but it's far easier to build a planned community in proximity to Phoenix than in proximity to other major US metro regions. There's a reason for this, of course: the desert is hot and it lacks water. Nevertheless, the company Culdesac has designed a walkable community in Tempe and most of the possible landing spots for Telosa are in the desert west.

Correspondingly, the ultimate experiment in futureproofing human habitation also began in the American west. Biosphere 2 was initially an experiment with human habitation in space (a subject for the next chapter). In two separate efforts (1991-1993 and another in 1994), residents of Biosphere 2 attempted to live in a closed habitation with a self-sustaining environment and food supply. It makes little to no sense that we might want 5 million residents in a desert city, but it made sense to put an experiment like Biosphere 2 in the desert – living conditions in space are harsh, but sunlight can be plentiful.

Biosphere 2 was more than just an attempt to keep the air oxygenated and the food supply sufficient, it was also an experiment in human social dynamics. The participants had to work together in a closed environment, finding ways to overcome conflict and find common ground. Anyone familiar with a homeowners' association or the co-op boards of New York City apartment buildings knows that aligning everyone toward common goals is a nontrivial task, even when the stakes are low. So finding good social dynamics was part of Biosphere 2 just as it is part of the urbanproofing agenda. Even with a small community, the Biospherians often struggled to align their goals.

There may be an extent to which the size of a community has an automatic and predetermined effect on social dynamics. Anthropologist Robin Dunbar argues that brain size correlates to community size and our ability to connect well with others. Based on studies of brains and group dynamics in a variety of species, he estimates that human beings can form strong community bonds with only about 150 people at a time. In human beings, the way to overcome Dunbar's number is through rules and laws: these govern us even when social bonds cannot. So while Biosphere 2 may not have needed explicit regulations (in fact, it did), any urbanproofing community will need clear rules for governance.

As we saw above, urbanproofers are committed to new models of political economy and governance, though libertarian viewpoints seem predominant. Whether such a model can efficiently guide any social system, much less one that purports to restore human beings to an appropriate relation with nature, remains to be seen. Urbanproofers expect that out in the desert some Burning Man-infused sensibility of self-governance and communion with nature can also happen in urban design (I leave aside whether Burning Man *actually* accomplishes either of those goals).

The hustle for new legal systems goes beyond Silicon Valley and applies uniformly across urbanproofing efforts. Quite in keeping with the libertarian ethos of tech culture, Próspera claims to be a "startup city with a regulatory system designed for entrepreneurs to build better, cheaper, and faster than anywhere else in the world." Their job application site indicates an interest in "candidates from diverse industries and backgrounds who are eager to revolutionize how governance is delivered and cities are built." It should be obvious that the entire idea of "delivering" government reorients government from civil services to product placement. The ethos of new laws is thus an outgrowth of entrepreneurial tech culture, as opposed to the religious or secular political goals that previously governed the philosophy and implementation of governance.

NEOM will have both civil and economic laws that differ from those of its parent country, Saudi Arabia. Early in NEOM's development, the extent of this remains unclear, especially considering that in 2022 the Saudi government felt compelled to assert that NEOM would be "completely under Saudi Arabia's sovereignty and regulations." Clearly, NEOM cannot accomplish the ultimate goal that seasteading communities have – total separation from present-day national borders; but some degree of legal re-jiggering is at stake for the Saudis. In this relationship, "it's complicated," as the kids say.

The Saudi approach, like those elsewhere in the world, alleges that urbanproofing could help solve the climate crisis. Simultaneous to being a hub for technology and business, and establishing a new set of laws, NEOM aspires to be an "accelerator of planetary regeneration" in which "nature has to come first." The other projects described here, such as Telosa, California Forever, and Próspera all claim something similar. The entire concept of the fifteen-minute city, for example, responds to the energy costs of transportation. Although Próspera engages the least with environmental concerns, even its website claims to support "sustainable growth" and shows smiling people standing amongst plants.

A drawing with all those happy people inside of Próspera has the (deliberate?) result of excluding everyone outside it; but no city is an island. For the Biospherians, unforeseen and vigorous external politics involving funders, support staff, and others outside the habitat contributed to what was often seen as the project's failure. In the study of science and technology, however, such problems are part and parcel with developing new ideas and new devices. In one of his stranger books, for example, French thinker Bruno Latour details how converging and differentiating interests doomed the Aramis subway in 1980s Paris. For all that Latour speaks on behalf of the train, he also points toward the many individuals and groups with a stake in the train's success or failure. The same is true of Próspera, NEOM, or any other futureproofing city.

Virtual cities

All of the cities described above are virtual in the sense of being more like vaporware than software; but some urban designs live in completely virtual environments and still provide a place for future-proofing. I don't want to claim that the dreams of urbanproofers will never materialize – change is inevitable. But right now, the new urban experiences have far more publicity than population. In contrast, the online habitation of virtual worlds clearly offers both inspiration as a design space and an experience for urban living that many users find compelling.

While the wildly popular online platform *Roblox* has more users, *Second Life* offers a more interesting set of lessons for thinking about new urban experiences because it provides a broader sense of habitation rather than focusing on game play. The full scope of virtual worlds includes a vast number of videogame environments, but also non-game online social spaces, of which the leading example is *Second Life*. Despite the collapse of media attention around 2010, *Second Life* remains a hub for online activity and has a consistent user base. Many decades ago, the artist Andy Warhol was enchanted by how you, me, the President, and Liz Taylor could all get Cokes and all the Cokes were good. But for all that equality, a virtual Coke is just not the same. So, *Second Life* lost its public luster when major corporations had a hard time selling their products in-world. That may be a lesson learned by the urbanproofers as well. Given their focus on business needs, a focus that nearly destroyed *Second Life*, urban-proofing efforts may not provide the community- and person-centered experience that they promise (not to mention the environ-mental sustainability).

Built to bring Neal Stephenson's fictional Metaverse into being, *Second Life* provides a 3d environment where users walk around as avatars in the digital space. Between 2010 and 2015, several authors (myself included) published books highlighting interesting aspects of *Second Life*. Major contributions include Tom Boellstorff's *Coming of Age in Second Life* (an attempt to provide thick cultural description of life there) and Wagner James Au's *The Making of Second Life* (Au was the "embedded journalist" at Linden Lab, the parent company of the world). Beyond scholarship, there were journalists, social commentators, and, of course, science fiction.

In *Snow Crash*, Stephenson advanced our collective understanding of what a digital space could even be. Early efforts to envision life in a computer world were inflected by geometry, from the movie *Tron* to Vernor Vinge's Other Plane in *True Names* and William Gibson's Matrix in *Neuromancer*. Stephenson, however, envisioned a place fully inhabitable as a commercial and social space – his Street in the Metaverse is Las Vegas on Disney-produced steroids. Stephenson showed little to no interest in *Second Life* or other attempts to build the Metaverse; but, as I noted in *Virtually Sacred*, the creators of almost every virtual world, including *Second Life*, recognize a profound debt to *Snow Crash*, as do the many users whose in-world designs borrow on the book and its expansion of cyberspace.

Among the most obvious SL examples, Extropia Core was a science fiction environment that thrived from 2006 to 2009. Extropia Core looked like a techno-fantasy of futuristic geometric buildings and was maintained by a group of residents most of whom held – no surprise here – transhumanist and libertarian views. Both perspectives can be seen in the community's backstory, which read:

> In the early part of the 21st century, a group of forward looking individuals start a foundation for creating a new nation, a techno-utopia in which individuals are free to be whoever and whatever they want to be, even augmenting or recreating themselves technologically if they so choose.... They decide to create a floating city to build their dreams upon. They call their new nation Extropia.

The parallels to seasteading and Próspera are particularly striking here, though one sees hints of all this in the entire urbanproofing effort. There's a transhumanist enhancement approach to Próspera and at least some of the seasteading efforts. At the same time, the question of individual freedom emerges from techno-anarchist and libertarian trends in transhumanist thought just as it does in the urbanproofing offshoots of all these cultural trends.

Drawing on the possibility that new forms of living extend new options for regulation and governance, the *Second Life* region of Al Andalus offers a very different perspective on community design. Al Andalus, which I also describe in *Virtually Sacred*, sought to recapture the cosmopolitan worldview of Andalusian Spain. For centuries in the medieval era, Jews, Christians, and Muslims lived in relative harmony and vibrant intellectual exploration. While this ended with the expulsion of Muslims and Jews by Christian rulers in the 15th century, it stands as a hallmark of human potential. In its land covenant, the *Second Life* region of Al Andalus states:

> The Al Andalus Project is a Second Life attempt to reconstruct 13th Century Moor Alhambra and build around this virtual space a community of individuals willing to explore the modalities of inter-action between different languages, nationalities, religions and cultures. The principles upon which this project is founded include political participation, separation of powers, justice and the rule of law. Membership in the community is open to all, regardless of sim land ownership, second life premium status, species of avatar, gender, religion, national origin, sexual orientation or any other traditionally separatist classification, either real or apparent. The plan is to create a system of political and legal governance, based on notions of community self-governance, active citizen participation, equality, dignity, social justice, democracy and human rights. Idealistic? Absolutely. Possible? Let's find out.

There is little to directly connect the Al Andalus realm of *Second Life* with the urbanproofers except the use of digital technology and the desire to recreate human government. Of course, the Al Andalus participants did not believe in governance as a product for sale and delivery, nor did they advocate "disruption" or other Silicon Valley catchphrases. Instead, they reached into the archives of history to find a model for human cooperation. Interestingly, Al Andalus was supported by the Saudi oil company Aramco, which published a piece praising the community as a "metaphor for the future." It would certainly be interesting to know whether any line of influence traces directly from *Second Life* to NEOM.

Neither Extropia Core nor Al Andalus remains today. Al Andalus was intended to run as a two-year experiment but due to the enthusiasm of its members it lasted five (from 2007 to 2012). While *Second Life* remains a strong and profitable virtual world, it is hard for any community there to last for long. The cost of doing business in *Second Life* means that without substantial community support or a patron willing to throw thousands of dollars per year to maintain a region, eventually it will close.

It may be a valuable lesson to the urbanproofers that early enthusiasm cannot sustain a community forever, especially when the costs are high. The musician Akon announced a high-profile multibillion-dollar future city in Senegal...a project leaving nothing but a partially built welcome center when it collapsed in 2025. Perhaps some version of Akon City will rise along Senegal's coast; but it certainly won't be the futurist paradise he promised in 2018. Calvino describes cities that dream, and it is those dreams more than the dreams of their founders that make them real.

Philip Rosedale, founder of *Second Life*, once said of the virtual world that "it'll breathe by itself, if it's big enough. We're helping because we're going in as avatars. It's simply the fact that if the system is big enough and has enough complexity, it will emerge with all these properties. People come from out of the dust." But should we come from the dust, as we read in the Book of Genesis, then likewise unto dust we shall return. Like Biosphere 2, *Second Life* communities were, ultimately, experimental. Perhaps with care, lessons can be gleaned that will help urbanproofers design communities and environments that are economically sustainable. But such sustainability may not simply assume a commensurate social harmony, low energy impact, and/or communion with nature.

Cities of earth and sky

Digital communities disappear into the ether, but physical spaces always leave a mark. There are ghost towns from America's past, for example, places where the buildings remain but the people do not. What's left of Hadrian's Wall stretches across England (most of its stones now serve roads and buildings) and the Vijayanagara empire left the carved city of Hampi. Some of our historical legacies are empty of all but tourist vendors. There are also places haunted by their pasts but where people still cling to the landscapes they've always known: post-industrial and post-mining towns where industry disappeared and economic opportunity left with it. There are other places whose religious inspirations remain archived in the names of streets and neighborhoods, but which would be unrecognizable to the Oneidans, Fourierists, Shakers, and other religious and social reformers who founded them. The urbanproofers take up the legacy of these communities, combine it with digital technology, and attempt a total reformation of what it means to live in a city. Their futurist vision is, like those religious reformers, one of salvation: they believe they are recreating urban communities as a response to the dangers faced by our species.

The urbanproofers don't generally use the language of existential risk, though key players in urbanproofing also play a role in other futureproofing efforts. Rather, they skirt around the question of what the future will be, presenting themselves as uniquely capable of creating the forms of governance, social experience, and environmental communion that humanity both craves and needs. Whether that rhetoric is actually more important to them than creating new modes of revenue generation is an open question. But whether these "future cities" are just new forms of economic oppression or a genuine path forward, they are part of a worldview that generates public controversy and enthusiasm alike.

Calvino's book drives us back to our shared goals for cities and for the future of humanity. "Cities also believe they are the work of the mind or of chance, but neither the one nor the other suffices to hold up their walls. You take delight not in a city's seven or seventy wonders, but in the answer it gives to a question of yours," says Marco Polo, to which the Khan replies: "Or the question it asks you, forcing you to answer." These urbanproofing cities purport to answer the question of humanity's future. But it's likely that their imagination actually conjures some other question, one about the role of power and money. Perhaps their walls are sufficiently robust to ask and answer many questions, and thus we can see a series of byways and buildings that genuinely move toward a better, more humane future.

In "Tower of Babylon," science fiction author Ted Chiang reworks the Biblical story of Babel into one of devotion rather than hubris, but simultaneously to reveal the possible emptiness of our transcendent dreams. Rather than marking human pride, Chiang's tower is a relentless pursuit of union, of the human desire to connect with the divine. But when they finally break through to Heaven (leading to a torrential flow of water reminiscent of God opening the "windows of the heavens" to produce Noah's flood), one character succeeds in reaching up and out of the sky into what lies beyond. He crawls through a cave, only to find himself back on the surface of the Earth. Chiang speaks of Babel as an attempt to get closer to God in a way that lands people right back on Earth where they started. Is there, perhaps, nowhere to transcend?

While most of the proposed "cities of the future" involve carving a glorious new kind of city out of wilderness, at least one seems grounded in existing infrastructure: Brownsville, Texas. No one has yet advocated that Brownsville be transmogrified into a 15-minute city or suggested that forestland could weave through it and connect users to nature. Nor, indeed, has even its most famous tenant, Elon Musk, suggested that it become energy independent. But Brownsville is the home of Musk's SpaceX, so it probably deserves mention in an account of urban futureproofing. After all, the lessons learned on Earth are relevant to Musk's vision of the future.

Though plans are still amorphous, Musk's vision extends beyond the borders of Brownsville. Late in 2024, SpaceX employees filed a petition with the state of Texas to incorporate a nearby town, Starbase. One of the town's residents, whose husband works for SpaceX, hopes that Starbase could be "a model of what new cities should be." At present, that model points only toward what journalist David Goodman notes is a return to the past: a company town. SpaceX owns nearly all of the land in what would be converted from unincorporated desert to incorporated city. SpaceX also handles most medical and educational facilities; so likely the shift would demand some form of state assistance and remove cost from the company's ledgers. Money aside, it took only a matter of months before SpaceX began a re-zoning process based on eminent domain: residents were told they might be forced to leave their homes in 2025. I will note criticisms of SpaceX in chapter six. For the moment, it's just worth saying that there is legitimate question as to what kind of city Starbase will be. In what fashion SpaceX, Elon Musk, or the residents see Starbase becoming the future of urban design remains murky; but for some people the pursuit of human spaceflight legitimates that future.

That future has not been cut from whole cloth in the 21st century; after all, the "Space Age" already happened. In the middle of the 20th century, as the Cold War loomed and the aerospace and defense industries boomed, art and architecture celebrated our dream of reaching the stars. From the swooping curves of Eero Saarinen's TWA terminal at John F. Kennedy International Airport in New York to William Pereira's ziggurat-like headquarters for Rockwell's Autonetics division in California, designers reached toward the skies and spoke of transcendence in the past and present. It's not all that far from these to Spaceport America's "Gateway to Space," built to launch Virgin Galactic's suborbital tourists in New Mexico. Such buildings stand as testaments to the intersection of technology, transcendent aspirations, and the power of design to shape our perspective.

Like genetic engineering, robotics, and AI, city design is part of the human pursuit of earthly transcendence. So much so that questions of urban life resurface when we build rockets and imagine a future beyond the atmosphere. Cities on Mars, space stations, interstellar colony ships: all of these draw on the braided legacies of transhumanism, science, literature, and urban design. They are part of urbanproofing, but even more they are part of the human expansion beyond Earth, a dream of our transcendence.

CHAPTER FIVE
THE ETHEREAL AGE

The sky has been long the domain of gods, and thus the final aim of our technology. Soaring cathedrals and the wings of Icarus. When my wife's great-grandfather flew a Pan-Am airplane on December 1, 1958, he received a certificate noting that here, in the "Domain of Phoebus Apollo Ruler of the Sun and Heavens," he gained a wondrous new status. The certificate, a keepsake of the family, says that he, "once earthbound and time-laden, is now declared a subject of the Realm of the Sun and of the Heavens, with the freedom of our Sacred Eagle...That with the speed of Our Winged Chariot this subject did fly the Pacific Skies over the International Dateline, which mortals designed to mark off in the limit of days Our Eternal Course through the Skies...That by so crossing this divider of days, the Today of mortals at once becomes Tomorrow and all is confusion...That this subject is commanded to hold ever close this Celestial Decree so that in the final account of earthly days, the balance will stand true forever more."

The certificate's divine reckoning of space and time unveils important aspects of the futureproofing effort. We cannot clearly glimpse the path from today to tomorrow and suffer the impossibility of predicting where our technologies will lead, but Moravec and

Kurzweil argue that even the end of the universe cannot end the cosmic transcendence of our digital mind children. In this way, the futureproofers reject the weight of time and, through their galactic intentions, space as well. Our dreams have long rejected our earthbound existence, and the future of our species must ultimately rest on a technological ascent to the heavens.

It is likely that for as long as human beings have thought about gods, angels, and other transcendent creatures, we have looked for them in the sky. Not exclusively, of course: human beings have always seen magic in the earth and sea as well. But the sky rises above us, providing the nourishment of rain, and startling us with the glory of the Sun, nearby planets, distant stars, intermittent comets, and meteor showers. Just as we, today, cannot help but thrill at the mystery and wonder of it, human beings and our evolutionary forebears have probably done so since time immemorial. It is no surprise, then, that traditional religions often envision the salvation of humankind in the stars, nor that the futureproofing religion does so as well. All of the technologies at stake in futureproofing cohere around spaceflight. Existential risk, on the other hand, is what makes spaceproofing necessary: if we remain earthbound, humanity cannot remain safe forever.

The existential risks to our species include many things human beings can, in theory, control; but they also include inevitabilities beyond our influence. If we cared to, we could stop runaway climate change and we could limit the use of genetic engineering or the advancement of AI. But we cannot control supervolcanoes, asteroids, or the sun going supernova. Even if, by some miracle, we could actually create a planetary defense system that protects us from extinction-level asteroids, it seems out of the bounds of our technological future to control the seismic activity of the Earth; and there is no reason whatsoever to believe we could control the lifecycle of the Sun. Sooner or later we are in trouble, which surely explains why Blue Origin (Amazon founder Jeff Bezos's space company) has an emptied hourglass in its company crest.

If we take a sufficiently long view, there is no future for humanity unless the species disperses across space. Only if there were enough people scattered across planets, solar systems, and even galaxies could we presume that human life, thought, and culture will outlast any and all dangers. Moravec and Kurzweil even go so far as to suggest that a sufficiently interstellar, transcendent posthumanity can prevent the end of the universe! While this last is hyperbole of the first caliber, it remains true that the annual probability of all human beings dying on Earth is greater than zero (even if small). That means such extinction is, ultimately, inevitable. After all, many species have come and gone on our planet; unless we act to avoid it, we will be among that number.

As with other technological dreams, there is a cottage industry of pop culture that hopes for interstellar protection from extinction. A full description of existential risk occurred in transhumanist communities long committed to human transcendence, and a conjunction of spaceflight advocacy with existential risk parallels the popularization of those risks in discussions of AI, urban design, and genetic engineering. The 2024 announcement – and, more importantly, the fact that this was news – that the production firm Legendary TV purchased the rights to Neal Stephenson's 2016 award-winning novel, *Seveneves* (a story of Earth's near-destruction and humanity's escape), reflects both the ongoing ability of Stephenson to predict the direction of public interest but also the widespread engagement with these ideas. After all, it was the threat of an asteroid hitting the Earth that Neil deGrasse Tyson used to defend astrophysics while eating spicy chicken wings on the hit YouTube show *Hot Ones*.

All through this book, it has been apparent that existential risk is at the heart of transforming a loose ideology or futuristic perspective into a modern mythology of human transcendence. All the technologies discussed, from urban design to AI uploads, have bounced around tech culture for decades. But the dread provoked by existential threat motivates people to see these technologies in a new light. What were once fringe perspectives now form the core of bestselling books and provoke new forms of corporate culture.

As we gaze upon the stars, wondering if they could be our ultimate destination, we are hurtling into new economic systems, new visions of urban reality, and new iterations of the human condition. Spaceflight is the location where all of our technologies converge in the futureproofing of human civilization. To make that possible, spaceproofers leverage existential risk alongside traditional religious goals to build up and sustain our technological pursuit of human (or posthuman) salvation, our journey to join the cosmic glories above.

Nothing new beyond the sun

We have long seen the stars as the home of gods and angels; so the contemporary dream of transcendent humanity in the stars is nothing new. But in the 21st century, we know that we can leave Earth orbit, and we suspect we could find a way to take up residence on planets, asteroids, and moons. Once upon a time, we expected only death could get us there; now, it is precisely the fear of death that drives us upward.

In the Abrahamic traditions, for example, there are repeated references to "God above" and to the angels as heavenly stars. The ancient Jewish text 2 Baruch promises that when the saved are "glorified in changes" that they will "be made equal to the stars." The Christian evangelist, Paul of Tarsus, describes heavenly bodies as possessing "splendor" (1 Cor 15:40-41) and then proceeds to discuss post-resurrection humanity as taking on imperishability, glory, power, and a spiritual body (1 Cor 15:42-44). "And just as we have borne the image of the earthly man," Paul writes, "so shall we bear the image of the heavenly man" (1 Cor 15:49).

Not every religion makes such direct correspondence between human salvation and the stars; but there are similarities from beyond Judaism and Christianity. For example, David White makes a similar case regarding ancient Indian yogis, who would project themselves through the sun to attain the land of immortality. In Hindu thought, the Ganges River is not just earthly, but interstellar: it is the Milky Way. And when the dead briefly emerge from it toward the end of the ancient epic *Mahabharata* they arrive in glorified form. In other cultures and mythologies (not a word I use pejoratively), the Sun, Moon, and planets get associated with divine beings. Heroes get deified and made into constellations. There is simply an obvious connection between things that glow in the sky above us and our imagination of the divine and the mythological.

In the 20th century, a cottage industry in UFO religion has carried out this legacy. Groups like the Raelians substitute alien entities for the divine, explain human achievements like the pyramids with recourse to those aliens, and anticipate various ways of joining the interstellar company. Occasionally, this takes (presumably) disastrous course, as when the Heaven's Gate community committed mass suicide so as to elevate their spirits to the Comet Hale-Bopp in the year 1997. But usually such movements are simply harmless repetitions of the ongoing pursuit of heaven.

When we speak of human salvation in the stars, of accompanying divine beings in their celestial dance, we speak of nothing new. But sometimes that pursuit gets reconfigured in new ways. In his musical intervention, for example, the experimental jazz artist Sun Ra pursued an "altered destiny" that exists "on the other side of time." In his movie, *Space Is the Place*, he says "if the planet takes hold of an altered destiny there is hope for everyone. But otherwise, the death sentence upon this planet still stands. Everyone must die." While his concern is spiritual death at least as much as literal, he says "I will take you to outer unseen worlds" in a promise of new life. As "the myth talking to you," Sun Ra and his music promise what the film labels "tomorrow's realm...another world" as his spaceship departs Earth. Such has been the dream of countless human beings as they gazed skyward in astonishment at the wonders above. So if humanity has an angelic transcendence in its future, this will develop through its destiny to reach the stars.

Destination: arisen

Religion motivated people to dream of lives and worlds in the stars, but those dreams were metaphorical until the dawn of the 20th century. The technological confluence of telescopes, high energy projectiles, electricity, motorized engines, and other breakthroughs made the unearthly appear tangible. As early as 1608 (though not published until after his death), Johannes Kepler could dream of a trip to the moon, with human beings reaching it through witchcraft and finding a new species of beings living there. While his story was unlikely to have been taken seriously by more than a few technoenthusiasts or cryptoreligious conspiracists, it marked a watershed moment in humanity: Kepler proposed that we *physically* enter the world of angels and gods.

Kepler's daemonic journey, like Jules Verne's later moon cannon, was obviously pure fantasy; but concrete technological efforts began making spaceflight more plausible long before we would actually leave the atmosphere. An early and influential champion of such spaceflight was a man familiar from previous chapters: Nikolai Fedorov. Fedorov lived a life so eccentric that his own friends sought and failed to bring him into a more comfortable mainstream. A lifelong bachelor, Fedorov lived in voluntary poverty, giving away almost all of his money and possessions. He apparently refused a bed and slept on the one luggage trunk he owned, rejected fine foods in favor of basic staples, and even refused to wear a coat in the wintertime. As a librarian, he came into contact with many of Moscow's elite, and, as George Young tells the story in *The Russian Cosmists*, he was a valued conversationalist among the intelligentsia, such as the literary circle that included Tolstoy and Dostoevsky. These latter even conspired to improve his living conditions, but any luxuries they provided – from a bed to better food – were swiftly refused or given away. Whether in spite of or because of his legendary asceticism, Fedorov earned the respect of intellectual Russians in the guidance he offered to library patrons.

A moderate number of Fedorov's writings survive, thanks largely to the prompting of his friends who wished to see him attain greater notoriety and even strove to publish the work for him. Despite their efforts, Fedorov's writing went underground during the Soviet regime, to resurface in the attics of old homes upon the collapse of that government. Fedorov's impact, however, went far beyond his writing. He changed the lives of important people – or people who would come to be important – through his personal relationships. It was Federov, notes Young, who noticed the prodigious potential of a nearly deaf young man, largely uneducated and without prospects. That young man, Konstantin Tsiolkovsky, would become one of the great theorists of the Russian space program. Not just Tsiolkovsky, but a host of scientists, poets, and literary authors learned from Fedorov. He thus sent shockwaves into history, including western Europe, as some of his associates ended in up in 1920s Paris, where they'd likely have met, and perhaps influenced, European futurists and transhumanists like Julian Huxley and Pierre Teilhard de Chardin.

The key to Fedorov's philosophy was the "Common Task": the pursuit of immortality and resurrection of the dead. As a Russian Orthodox Christian, he devoutly believed in the salvation promised in the gospels; but he considered it human obligation to realize those promises. The end of war, the establishment of universal kinship, and the delivery from death were key aspects of his thought, and all point toward his vision of universal immortality. Much of this does seem widely shared among people, but there were some peculiar aspects of Common Task. In particular, Fedorov believed truly the Biblical god's dictum "from dust did thou come and to dust thou shalt return," and, further, that the dust of our ancestors would have dispersed throughout space. So, to resurrect "the Fathers," it behooves us to explore the solar system and collect their dust.

In his brief essay, "Astronomy and Architecture," Fedorov suggests harnessing the power of the sun using zeppelins to capture and broadcast the sun's energy to Earth, and then using this collective power to literally take control of Earth's movement through space. He referred to our planet, controlled in such a fashion, as an "electric boat;" using it we should soar through space to vacuum up the dust of our ancestors as part of a resurrection program going all the way back to the biblical Adam.

Strange as this theoretical enterprise might seem, it was crucial in developing the Russian space sensibility. It provided an imaginative energy for what might otherwise seem impossible, a mythical force to justify seeking spaceflight. Perhaps this was why Tsiolkovsky found his way to the calculations necessary for departing Earth gravity, opening the door to Sputnik and beyond. Tsiolkovsky's own plans for Earth included geoengineering projects that hindsight clearly shows to be disastrous (for example, he favored deliberate global warming to supposedly enhance farm output), and he was committed to expanding humanity to the stars.

Perhaps not a direct follower of Fedorov, the Russian thinker Alexander Bogdanov wrote science fiction about enlightened communism among spacefaring humanity that both affirms and contradicts Fedorov's vision. His short story, "Immortality Day," describes a time when mortality has long since ended, though with it also the purpose of life, according to the story's protagonist. This is, perhaps, a curious end of tale for the first person to attempt blood transfusions to regain his youthful vigor. But George Young notes that Bogdanov remains always committed to Earth: any movement to Mars would be always respectful of the needs of humanity here. Perhaps this explains both his cynicism regarding immortality and highlights the difference between Bogdanov's perspective and Fedorov's (and thus of later Cosmists and transhumanists also). For so many thinkers, spaceflight is about transcendence, and accepting the limits of human mortality would have prohibited complete commitment to leaving the Earth.

A crucial juncture appears in Fedorov's work, one that would eventually come to flow through Euro-American conversations about spaceflight. For whatever reason of accident, Fedorov seems to have been the first to see concrete connections between altering humanity, ending mortality, resurrecting the dead, and travel beyond Earth's orbit. His efforts, both through the library and his publications, drove a new Russian myth.

That same myth eventually had corollaries in American views of spaceflight, though these took somewhat longer to percolate. Importantly, however, they held greater public significance than Fedorov's.

While the latter inspired a few elites, the American vision became a public urge to fulfill the nation's destiny. Eventually, the American faith in humankind's interstellar destiny adopted a religious view akin to Fedorov's. It began as a collective pursuit of American mythological ascendance and became a key locus for dreams of immortality and personal transcendence.

In her book, *Destined for the Stars*, Catherine Newell describes the multipronged effort to create an American myth for space exploration. From Chesley Bonestell's paintings of little astronauts standing on Saturn's moons to Disney's 1950s television specials and construction of Tomorrowland to eventually the launching of the Saturn rockets and the Apollo missions, a confluence of factors recreated the American frontier myth in space. After colonizing Americans previously saw themselves enacting "manifest destiny" in their conquest of the central and western regions of N. America (an effort also tied up in technologies, such as plows, telegraphs, and trains), mid-20th century Americans went looking for a new world to take as their own. The "final frontier," as Gene Roddenberry named it in *Star Trek*, was the open universe for American expansion.

It was the definition of space as a frontier that made America's moon landing possible. While Cold War politics certainly contributed to American expenditures on the Space Race, Sputnik did not launch until 1957. By that time, Americans were already watching futurist dreams of space on TV and reading about them in influential magazines like *Collier's* and *Life*. As Newell shows, it wasn't so much fear of the Soviets so much as love of a new frontier that generated American enthusiasm for spaceflight. Illustrations and entertainment cultivated a renewed sense of identity, one borrowed from the national myth of conquering the west.

The sacred myth of American expansion resonates with space expansion for material, political, and economic reasons. Euro-Americans declared that people needed more land and more resources, so they took these from indigenous Americans. But there are no indigenous Martians from whom to take that planet. Robert Zubrin, one of the longest standing advocates of colonizing Mars, says that doing so will meet our need for the challenging frontier but also ensure the safety of human beings on Earth. Rather than pit human against human in a death spiral of conflict over resources, he suggests that peace on Earth could be the outcome of colonies on Mars. He sees space exploration as a proper recognition that "each new life is a gift, every race or nation is fundamentally the friend of every other race or nation."

During the heyday of American enthusiasm for space, transhumanists and futurists were leading advocates for spaceflight. In the early 1970s, FM-2030 (still writing under his given name of Fereidoun Esfandiary) wrote that "this is a beautiful moment in human evolution. It is the stage at which the human species has broken away from its confinement to this planet. Suddenly we are no longer of this world only. We are transcending our world." The final page of his book, *Optimism One*, includes the revolutionary cry:

We now want cosmic rights.
We want the freedom to roam the universe.
We want nothing less than the right to determine our own evolution.
We want the right to live forever.
So long as we have not overthrown the tyranny death no one is free.

For FM, like for Fedorov, spaceflight links to immortality and the posthuman evolution of our species. Unsurprisingly, his associate Max More included "boundless expansion" among his Principles of Extropy, partially qualifying that as "expanding into the universe and advancing without end." The mid-20th century transhumanists certainly equated the space frontier with the evolutionary frontier and the salvation of humankind.

Despite such enthusiasm, the American space frontier myth collapsed under the onslaught of technological fears and a subsequent decline in civilian faith (and funding). Although Kendrick Oliver points toward somewhat anemic public support of the post-

Apollo spaceflights, the Challenger Disaster of January 28, 1986 seems to be the real inflection point in the American space program. Millions of Americans, among them school children, watched as the Challenger exploded ninety seconds from launch. The flight was especially well-publicized because of the inclusion of Christa McAuliffe, a public-school teacher; and so a generation of young people watched their collective dreams go up in smoke. The American public soon lost its taste for spaceflight, and Congress found reason after reason for restricting NASA funding.

The negativity around the disaster was no doubt exacerbated by a lingering confusion over NASA's goals: after all, no rockets capable of reaching the moon were in operation and there was no obvious effort to get back or beyond it. It's hard to maintain a frontier myth if you never leave your porch. After the reentry explosion of space shuttle Columbia in 2003, the clamor for spaceflight had settled into a dull roar, impossible to hear outside the circles of science fiction and space aficionados. The public cheered as Mars rovers overcame crises of dust storms, the challenging landscape, and technological hiccups while sending photographs of the planet's surface. But it was the plucky spirit of anthropomorphized robots that excited people, not the prospect of human beings following suit.

Curiously, it was in the midst of this public malaise that tech billionaires turned their attention to spaceflight. They threw their financial weight behind space travel just as governments diminished their own. The X Prize (later renamed the Ansari X Prize) was announced in 1996 to encourage commercial space travel, and it helped inspire the coming shifts in space exploration. The Paul Allen funded project, SpaceShipOne, won the prize in 2004.

While there could be only one winner, the billionaires lined up their investments and committed to space tourism en route to space habitations. Jeff Bezos founded Blue Origin in 2000, Elon Musk founded SpaceX in 2002, and Richard Branson founded Virgin Galactic in 2004. Branson began a collaboration with Scaled Composites, the company that built SpaceShipOne, while Bezos and Musk used their own companies to design and build rockets and related technologies. Space aficionados, business advocates, and technological cheerleaders swiftly heralded these efforts as an alternative to the old NASA approach; the new version of space exploration was thus dubbed NewSpace.

A New Heavens in a NewSpace

The post-Challenger decline in scientific trust and the compelling myth of spaceflight came with commensurate absence of investment in space technologies; but a renewed sense of cosmic adventure – brought on by futurist religious speculation and accelerated by public dread – produced a new industrial effort to reach space. The NewSpace movement in the United States is a conglomeration of public-private economic partnerships, cultural advocacy for the human future, international competition for resources and cultural supremacy, and a response to existential risk.

The space industry has always been something of a public-private collaboration. NASA needed contractors like Boeing to produce things like the Apollo rockets. NASA provided science, coordination, and funding. Boeing and other companies engineered solutions. It's simply not possible to produce something as complex as a human flight to the Moon or an orbital space station without the expertise of industry and government combined.

But the 21st century NewSpace movement takes public-private partnership and adds several new dimensions. Government, by and large, simply provides tax breaks and a general structure of goals and incentives. Private industry takes these, combines them with commercial goods (e.g., suborbital flights for the wealthy) and packages them up in *vision*. NewSpace is not just about getting US astronauts to the International Space Station (a legitimate problem once the shuttle fleet was grounded in 2003). It simultaneously enabled William Shatner's actual launch into the final frontier and an opportunity for the wealthy to re-envision their place in the cosmos. For example, when Virgin Galactic began taking pre-orders in 2004, it garnered hundreds of respondents and more than ten million dollars in contracts by early 2006 (with tens of millions more in the years to come). Companies like Virgin Galactic, SpaceX, and Blue Origin now compete with the contractors of previous eras, like Boeing and Lockheed-Martin. In 2024, the US Space Force announced that billions of dollars in contracts would be awarded to NewSpace companies. This push toward entrepreneurial spaceflight thus shapes a space industry that serves government needs, provides spaceflight to wealthy individuals, launches satellites, returns (perhaps eventually) investor value, and provides grist for the pro-spaceflight cultural mill.

Many people remember spaceflight triumphs with great fondness (I confess that I own a newspaper cover page from the 1968 moon landing) and believe the NewSpace movement will play a critical role in the future of human culture. Shortly before leaving my previous institution, I dropped by the college archives to visit my friend there. My plan was a friendly goodbye, but I couldn't help noticing the latest acquisition: someone had delivered a pile of newspapers and magazines heralding the "giant leap for mankind." For a generation of Americans, July 20, 1969 is vividly remembered. Wyn Wachorst, hired by NASA to provide a new "myth" for its mission, sees that landing as the signature event of the 20th century (an optimistic gloss on a century known also for the bombing of Hiroshima and Nagasaki and the twelve million murdered in German concentration camps). Wachorst, in keeping with his role, sees science as a spiritual quest and experiences deep nostalgia for the height of the space craze.

Alongside these flights down memory lane are hopeful glimpses of the future. NewSpace advocates believe that their technologies will solve the economic troubles of the modern era. In particular, they promote longstanding dreams that by moving advanced industry and mining into space we ensure what would otherwise be unsustainable economic growth.

In particular, investment in spaceflight could help us respond to the environmental costs that accompany our appetite for 21st century technologies. For decades, we have known that valuable minerals compose many asteroids; and the protagonists of our previous chapters, such as Hans Moravec, have encouraged us to consider how we can access these. Moravec argued that creating a space mining industry was one of the key benefits of robotics: by mining in space, we avoid all the environmental costs of doing so on Earth and can simply pipe down the resources (such as by use of a space elevator – a theoretically plausible but thus far impossible way of using a tether to move objects to and from a station in geosynchronous orbit). Moravec hoped the promise of space mining would be a catalyst for robotics investment, and the same resource competition drives the overall spaceflight industry. There can be little argument that we are in a resource crisis, in which both the extraction and refinement of earthly minerals lead to pollution, climate change, and awful human labor practices. Cost effective means of reaching space and extracting minerals from asteroids or the Moon could help save the Earth from the capitalist practices consuming it.

There are also political drivers at stake in 21st century spaceflight. It is well-known that Cold War politics contributed to the American and Soviet space programs. Now, there are international concerns over the militarization of space and the forms of cultural supremacy that come from spacefaring achievements. In 2023, India became just the fourth nation to achieve a soft landing on the moon and did so at considerably less cost than previous nations. They hope their Chandrayaan-3 lander is just the beginning of greater international visibility. Meanwhile, China, the US, and Russia remain competitors despite their collaboration on the International Space Station. These nations, and others who wish to join the Moon/Mars competition, all vie for the recognition that comes from successful landings. Each such landing is an opportunity to celebrate cultural heritage and declare one's nation to be at the forefront of humanity and its future. Scientific success acts as a validation of the political order.

While nations remain key players in 21st century spaceflight technology, the NewSpace entrepreneurs muddy the politics. The leading companies are western, but these companies do not necessarily serve national interests. In an essay published by the online magazine *Sapiens*, anthropologist Kimberley McKinson notes that customers of

Elon Musk's Starlink satellite internet service sign an agreement that "for services provided on Mars, or in transit to Mars via Starship or other spacecraft, the parties recognize Mars as a free planet and that no Earth-based government has authority or sovereignty over Martian activities." Starlink thus connects to Musk's spacefaring goals while affirming political as well as terrestrial transcendence. Musk wants us to leave our earthly nations behind as we travel to the stars. Perhaps that explains his chainsaw approach to the US government early in 2025.

In a 2021 interview with Peter Diamandis, Musk described his own vision of our spacefaring future both in terms of its sacrificial requirements and its historical necessity. He acknowledged "a bunch of people will probably die in the beginning" but argues it's all worthwhile. "Humanity is the agent of life and we have an obligation to ensure the creatures of Earth continue even if there is a calamity on Earth, whether it is man-made or a natural calamity – if you look at the fossil record there are many mass extinctions. It's about ensuring we pass that threshold where it is self-sustaining if some calamity prevents the ships from going there." He thus articulates the clear engagement between spaceflight, the NewSpace movement, and the religion of existential risk. For good or ill, the movement of SpaceX and others in this domain certainly opens the window of exploration and understanding: in 2024, a billionaire-financed SpaceX flight flew to the highest Earth orbit position in history and the crew engaged in a spacewalk wearing spacesuits that apparently surpass those designed by NASA and other government agencies.

In a 2019 lecture to the National Space Society, Jeff Bezos declared space exploration to be the answer to long range problems, such as the ultimate limits on Earth's energy. He does not conjure the spectre of total extinction, but runaway energy needs cannot be separated from resource extraction, climate change, and social conflict. As such, while Bezos simply borrows from Gerard O'Neill in his worries about energy access, the relevance of energy futures is contingent upon other, equally or more dark, future speculations (more on O'Neill and Bezos's debt to him in a moment).

Bezos is one voice in a conversation happening largely in the land of private entrepreneurship. It's hard to measure government enthusiasm for investing in space despite the rhetorical advantage to be gained by landing on Mars or the economic gains that asteroid mining would create. Aside from NASA, the US government, for example, prefers to let wealthy tourists create the opportunity for broader policy. Virgin Galactic, Blue Origin, SpaceX: these companies revolve around a supposed future of space tourism and meanwhile use government contracts to sustain operations while they try to create safe flying conditions. As astrophysicist and space enthusiast Michio Kaku says, "the political and economic circumstances are changing. A new cast of characters is taking center stage. Daring astronauts are being replaced by dashing billionaire entrepreneurs. New ideas, new energy, and new funding are driving this renaissance. But can this combination of private funds and government financing pave the way to the heavens?"

In a 2024 essay, journalist Jeannine Mancini described a 2014 TED conference interview between Charlie Rose and Google co-founder Larry Page in which feelings like Kaku's emerged. Rose believed Page once claimed he might will his fortune to Musk because he "had confidence [Musk] would change future." Page replied to Rose with "he [Musk] actually wanted to go to Mars – he wants to go to Mars to back up humanity...It's a company and it's philanthropical. So I think, you know, we aim to do kind of similar things." Spaceproofing seeks profit and salvation.

With the election of Donald Trump in 2024, the influence of NewSpace pioneers seemed likely to accelerate. Musk was a high-profile advisor and supporter during the election and was slated to do significant work on "government efficiencies." He was poised for unprecedented engagement with American space projects, from human flight to satellite deployment. Before the year ended, Trump tapped Jared Isaacman, one of Musk's spaceflight collaborators, to run NASA. The Musk/Trump fallout in 2025, however, grounded that potential: Musk and Trump took to sniping at one another on Twitter and the latter terminated Isaacman's nomination. In November of

2025, however, Trump renominated Isaacman, thus reigniting the spaceproofing agenda at NASA.

Neither Musk nor Isaacman can be finished off, as they both have too much money to ignore. Isaacman is the billionaire who participated in and paid the bill for the SpaceX high-orbit spacewalk earlier in 2024. He is thus an experienced space entrepreneur and space explorer poised to advance his and Musk's personal interests by aligning NASA goals with them. At the same time, he maintains the futureproofing line: in a December 4 post to Twitter/X, he vows that "Americans will walk on the Moon and Mars and in doing so, we will make life better here on Earth."

Moving beyond Earth is, of course, the actual plan for futureproofing humanity – and such planetary transcendence will not come easily. According to at least one group, the Hyperion Project, we can build interdisciplinary solutions to stellar expeditions but it requires powerful new modes of collaboration. The group launched a design project in late 2024, offering rewards for design solutions that promote social cohesion and project success for a multi-generational spacecraft attempting to reach a nearby solar system. To enter, teams needed an architect, an engineer, and a social scientist. The Hyperion Project's website lists the goal as designing "the habitat of a generation ship, including its architecture and society." The required cross-disciplinary conversation is among the more compelling aspects of the contest. Compared to the coding-only approach to AI ethics (which has devolved further and further into corporate control), the Hyperion Project at least dreams of a world where people work together across their specialties. Perhaps that effort will produce outcomes that assist the larger goal of human survival.

For decades, futurists have maintained that spaceflight is the only option for humanity. Among these, sociologist William Sims (Bill) Bainbridge played a key bridge role from mid-20th century space enthusiasm to 21st century futureproofing. Before he committed to that work, however, futurists like FM-2030 and Robert Ettinger enthusiastically endorsed spaceflight and – while they are not the subject of this book – so, too, did science fiction authors from Isaac Asimov to Roger Zelazny.

The mid-20th c. futurists certainly supposed that humanity's future lay in space, but their approach was not *explicitly* religious. Spurred by science fiction, Ettinger believed that humanity was at the cusp of universal wellbeing and biological immortality. Ettinger developed the idea of cryonic life suspension from reading a science fiction story, and thought such technology should be used to overcome death. He expanded on the idea of bringing an accidentally frozen person back to life by suggesting that we could deliberately freeze ourselves. If we are on the cusp of death, we have literally nothing to lose. Through whole-body or "neuro" (just the head) cryonic freezing, we could be suspended until such time as our resurrection became possible. But for Ettinger, the human adventure did not end at the borders of our atmosphere.

Ettinger, as noted in chapter one, was not really anti-religious any more than he was, precisely, religious; but his "Long View" saw humanity living immortal lives of fantastic potential in the stars. In *Man into Superman*, he asks, "can we settle for peace, faith, and virtue, and forego the stars?" and then answers, "my own belief is that we cannot." The intentional evolution of the species would mean new forms of bodies and all manner of new technologies built out of computation, genetics, and more. These would combine to such extent that even an individual's home would be part of their extended, cognitive universe (it would do some of the thinking, some of the nagging, some of the supporting for its inhabitant ... giving new meaning to Winston Churchill's adage "we shape our buildings, and afterward our buildings shape us").

Ettinger spends little time on spacefaring existence – choosing to focus on the changes to humanity itself – but that existence is a clear outcome for him. Imagining a post-scarcity future of technological mastery, he writes:

> One can build an idyllic picture of countless free spirits, each the
> owner of wealth to glut a million sultans, each a master of a genii of

> bottomless resources, each the king of his castle and the ruler of his
> realm, yet each a gypsy, not rooted in any turf, whose star-wagon can
> carry him over the gulfs of space as across the chasms of the mind.

Thus for Ettinger, the entire purpose of our technological gains will be to enrich human life in its entirety and permit a new, more cosmic version of our species.

In *Up-Wingers* (1973), FM-2030 writes that "the Space Age still in its infancy is catapulting us beyond the premises that govern life on this planet. We are witnessing today the very beginning of a cosmic dimension which is not only altering life on this planet but affecting our entire solar system and the universe beyond." FM consistently wrote of space technology and perceived in it the clear proof of a united and, eventually, transcendent species. To fly beyond Earth was to see it in its entirety, as a whole, and thus to reject the national and ethnic borders separating humanity. The spaceflight at the core of FM's thought reflected a pursuit of transcendent experiences and ways of living. "To limit ourselves to this infinitesimal speck in space is to limit our potential for cosmic growth. To transcend to a higher evolution we must transcend our earth-habitat." Beyond such limits, he hoped to travel deep into space and find places "bathed in perpetual sunlight and galaxy light of fantastic brilliance and beautiful colors the like of which we do not have here."

From his perspective, the universe provokes optimism but human religion and philosophy both resist it. FM argues that traditional modes of thought and life too often reject the world and humanity: they are pessimistic about human nature and convinced there is evil in the world. So, while Ettinger sees religious people as, perhaps, willing to join the Long View, FM thinks that religion has fundamentally failed humanity. To his mind, a philosophy of optimism is one that has left human religion behind.

But for all the devout atheism of 20th c. futurism, the same impulses that provoked ideas of transcendence and even the building of gods in AI infused dreams of spaceflight. The first person to present an argument for creating an explicit religion out of humanity's race to the stars was Bill Bainbridge, whose sociological research on the spaceflight industry led him to write the essays "Galactic Religion" and then "Galactic Religion 2.0." In essence, Bainbridge believes that humanity will fail to colonize space without a religion to motivate us. He recognizes the collapse of the American myth that had driven space exploration, because he argues that no real impetus remains. Drawing inspiration from Moravec, Bainbridge feels a new religion of immortal spacefaring humanity could drive us.

Bainbridge is a longstanding supporter of pattern identity and the idea that immortality, even resurrection, are possible through computer simulation of a person's identity. Early in the rise of the Internet, for example, Bainbridge argued that a sufficiently robust questionnaire could help produce a "mind file." And as a scholar of videogames (which was one of the many domains in which he produced seminal research), Bainbridge suggested that if his in-game avatar could act on its own in a manner identical to how he would have acted then the avatar represents a form of personal immortality. His galactic religion draws on these basic ideas to promise immortality to human beings.

If our information patterns are our selves, and replicating those patterns in new bodies provides us with immortality, then we could "beam" that information pattern to new planets and explore the universe. We could first send some form of mechanical system to new planets in traditional spacecrafts, and these could be programmed to assemble robotic bodies, habitats, etc. We could then move information at light speed using high-powered lasers, thereby sending our consciousness to the new bodies. Even if we never find a way to travel at faster-than-light speeds, we can slowly move our mechanical assembly systems from planet to planet, and chase after them at appropriate times and at lightspeed.

But immortality must be earned, right? Bainbridge follows traditional religions in this, arguing that there are no guarantees of salvation. Instead, anyone wishing for additional lives in space must contribute to the collective work of getting there. Not everyone needs to be an engineer, of course; there are many ways of contributing to human culture, and Bainbridge has an expansive view. We might honor a tremendous poet with additional life just as we might do so for someone who improves our rocket technologies. The essence is that we should be responsible for our collective human project.

Galactic responsibility and salvation in the stars is reminiscent of the cosmic office that Huxley ascribes to humanity. Huxley believed that since humanity was gaining power over its own evolutionary future then every member of the species has a tremendous responsibility, one that cannot be simply ignored. What Huxley labels our duty, Bainbridge turns into a promise of redemption. Care and contribute if you want to see the stars: you can earn your future lives.

Urbanstellar

If human beings are heading to the stars, we're going to need places for people to live, and that means cities. The same logic applied to the construction of futuristic cities in chapter four will remain active in space. Cities, being machines for living in, represent the kinds of living that might be possible. The space cities of the future have their antecedents in utopian urban design, as noted, but also in the 20th century aerospace industry. Especially in the designs of William Pereira, the architecture of California's aerospace buildings "remind us how an industry, a region, and an era were defined by the science and engineering of seemingly limitless possibilities," writes Stuart Leslie. And while he also notes that those same buildings "also remind us that nothing stays modern forever," the race to design space habitats is one that draws on presumptions of an infinite future for humanity or our inheritors.

In the same 2019 speech when he announced energy limits implied the necessity of colonizing space, Jeff Bezos tipped his hat to Gerard O'Neill, famous for having imagined immense space habitats with Earthlike living conditions in his book *The High Frontier*. As Mary-Jane Rubenstein notes, Bezos was a devotee of O'Neill during his undergraduate years at Princeton. So it's not surprising that in his current space advocacy, Bezos more-or-less just cribs O'Neill's position – all of it except the need to improve conditions for the poor (about which O'Neill, but not Bezos, was outspoken). Space habitations complement the planetary aspirations of the futurists who dream of living on Mars or the Moon, with these two options drawing on 20th century theories to become 21st century salvation.

Unsurprisingly, existential risk – as a phrase – was not yet threaded throughout the 20th century imagination of space cities but the concept was there. O'Neill did note that "soon after a civilization reaches our own modest level of technological competence it becomes unkillable in the physical sense; the reason is just the topic of this book: the movement of life into space." But for his logic, the causal direction goes the opposite way compared to spaceproofing. Space becomes possible, and therefore humanity becomes unkillable. But for the 21st century spaceproofers, it is precisely to make humanity unkillable that space becomes necessary.

Similarly, astrophysicist Jill Tartar has seen the search for extraterrestrial intelligence as evidence that humanity might survive its self-inflicted risks. Her work on signals from outer space is an attempt to find hope: "the simple detection of the signal will tell us one thing, and that is that it is possible to survive the technological adolescence that we seem to be going through." In particular, finding a nearby signal would be extra cause for optimism: "if we can find a signal easily when we first try, they have to be, on the average, fairly close to us. That means that there have to be a lot of civilizations, and this just can't occur if the average civilization only lives for a hundred years and then blows itself up. It's totally improbable that you could have civilizations that are detectable and close to you and easy to find if they, as a rule, only live the life span that we've had on this planet, and then do themselves in." Supporting evidence that self-extinction isn't a necessary outcome does not assure us of a future beyond the current century. But it would be helpful, and perhaps lends credence that we can, ourselves, expand beyond Earth.

The existential risk crowd includes advocates of both stations on Mars and stations mid-space. During the years O'Neill advocated floating habitats, others – like Robert Zubrin – pushed for stations on the Moon, en route to stations on Mars. These late 20th century visions provide scientific justification for science fiction futures. Governments and their private sector collaborators seem split between the two visions, with NASA promising $100million in 2024 to help develop new space stations (far more modest than O'Neill's vision) and simultaneously pushing toward a supposed moon base in the 2030s.

O'Neill's space habitats could, he proposed, be *more* Earthlike than stations on the Moon, Mars, or elsewhere. By spinning they could create artificial gravity at Earth-level, and with sufficient effort they could have entire ecosystems inside. Rather than terraforming a planet to be like Earth, O'Neill proposed we could have enormous rotating cylinders, spheres, or wheels. The rotation simulates gravity and, at least in theory, it would be possible to create an Earthlike environment and grow plant life. A transparent "ceiling" would allow for the experience of day and night at intervals based on the rotation of the habitat. O'Neill even saw the potential for space inhabitants moving from habitat to habitat, perhaps living in one and working or vacationing in another. These habitats could each maintain their own environmental conditions, allowing space colonists to hop from village to village, experiencing different cultural and ecological conditions.

Although we remain very far from space habitats, O'Neill believed that even with existing 1970s technology it would be possible to establish space colonies in a matter of one to two decades. Recognizing that his time frame was optimistic, he nevertheless claimed that we could have been mining the moon and building habitats – by our reckoning – quite some time ago. Like many dreaming urban-proofers, he even suggests that across habitats the governments could vary, and he also directly states that space habitats would be governed under corporate charters rather than nation-states. Given that kind of lineage, it's no wonder that 21st century tech billionaires endorse life beyond Earth.

From an engineering standpoint, however, creating massive rotating space habitats would likely be harder than building a station on the Moon or Mars (though certainly much easier than terraforming a planet to provide it with a breathable atmosphere, animal life, etc.). The engineering likelihood, but also the *vision* of planetary bases underlies the enthusiasm of both the US and the Chinese governments to envision a lunar station in the 2030s. The confirmation of large caverns on the Moon in 2024 adds to the potential for lunar bases, as such caverns would provide shelter from cosmic radiation, meteors, and other dangers.

The timeframe for designing and building such space stations is unclear, at best, especially given the drawn-out process of creating space stations like Skylab and the International Space Station. It is unclear whether we will meet this need prior to human- or chance-induced catastrophe; but NewSpace entrepreneurs are fundraising toward that end. By the mid-2020s, companies like Max Space and Bigelow Enterprises were hard at work producing expandable space habitats to use in orbit and in the lunar caverns.

These entrepreneurial visions follow on early scientific work to understand what it would mean to build space habitats. Skylab and the International Space Station are obvious examples of people living in space; but they do not purport toward developing self-sufficiency. In contrast, the Biosphere 2 project of the 1990s did exactly that. Despite the cut and thrust of politics that follows economic transactions in science and often wreaked havoc on the experiment, Biosphere 2 managed interesting engineering feats. For two years, its residents remained largely self-sufficient, caring for miniature versions of an ocean, a rain forest, a desert, and a savannah. They had farmland where they cultivated their own food, and they maintained a tight seal on the exchange of oxygen from the outside world. It wasn't perfect in any of these efforts, but Biosphere 2 certainly advanced our ability and our knowledge.

It seems that all rivers have at least a tributary in the Russian Cosmists, and Biosphere 2 is no exception. As Mark Nelson narrates in his excellent book, there were ex-Soviets who consulted on the project (because they had worked on a similar effort in their own country) and one of them, Russian biophysicist Josef Gitelson, "said he had only one serious criticism of the project, but an important one: it had the wrong name and should be called Noosphere 1, not Biosphere 2." A reference to the noosphere is a reference to the early 20th century Russian Cosmist, Vladimir Vernadsky, a contemporary of Pierre Teilhard de Chardin. For Vernadsky, the noosphere is the emergence of thought alongside biology, a convergence into one coherent, technological whole; Teilhard de Chardin, who may have been the originator of the term, included also the evolutionary convergence of human minds and spirituality with his god. Decades later, Nelson carried translations of Vernadsky's work even before a published version existed in English. So more than half a century after a Russian Cosmist and a French Jesuit contemplated the nature of cosmic evolution, Americans brought the noosphere with them to the construction of Biosphere 2.

Science fiction revels in extraterrestrial cities, drawing inspiration from both O'Neill and more commonplace planetary landscapes. We see variants on O'Neill's spinning habitats in movies like *2001: A Space Odyssey* and books like *Neuromancer*. We see planetary stations in *The Expanse* and *Foundation*. For decades, science fiction authors have presumed that it is our destiny to reach the stars. In this, they both draw on and contribute to the mythos described by Catherine Newell. Not every science fiction author supposes that we will leave the Earth, but most do. After all, with 100, 200, or 500 years of progress, shouldn't we have the capacity? And when do human beings refuse to do anything of which they are capable? We've gone from the *New York Times* mocking Robert Goddard in 1920 to a reality tv show called "Stars on Mars" in 2023 (Olympic figure skater Adam Rippon emerged victorious). Throughout that century, humanity has woven in and out of narratives about space. Science fiction authors participate in this and inspire the technological future.

"Stars on Mars" clearly borrows from Biosphere 2 and its effort to simulate the construction of extraplanetary colonies in Arizona. Biosphere 2 was a public spectacle of science, survival, and the pretense of spaceflight (Biosphere 1 is, of course, the Earth). Like a reality TV show, the experiments with people living in the Biosphere 2 habitat came with both political controversy and scientific struggles. The premise was building a self-sustaining ecosystem such as one might try to establish off-planet, and do so in a way that a group of human beings could live within that ecosystem. There were challenges getting enough nutrition (even with an intentional low-calorie diet), maintaining the chemical composition of the atmosphere, and perhaps most saliently, keeping everyone happy.

An important part of what became a debacle was the outside politics. Inside the habitat, people worked together despite the development of factions that disagreed over the ultimate direction of the project. But outside the project there were funding problems, debates over the right management of the habitat, and even a corporate takeover by then-finance specialist Steve Bannon (later to be convicted felon for contempt of Congress before being pardoned by Donald Trump). Given the influence of the tech industry and its billionaire investment competitors in spaceproofing, perhaps we should take a warning from the fiasco surrounding Biosphere 2.

While Biosphere 2 points toward the challenges, the idea of planetary habitats cannot be dismissed. Bezos has taken up O'Neill's mantle; his billionaire competitor, Elon Musk, remains committed to cities on Mars. We've seen already Musk's commitment to a cyborg future and of course he is known for his investment in technologies from advanced automobiles, "democratic" information flow in his dubious Grok AI (designed to share "news" by aggregating the most common statements on Twitter/X), brain-computer interfaces, and spaceflight. But his vision for space goes well beyond flights for the wealthy, satellite delivery for NASA, and his own Starlink communication system. In a conversation posted to YouTube by the XPRIZE Foundation in 2021, Musk asks "which comes first, a city on Mars or World War 3?" This simple question reveals both his investment in extraplanetary cities and the existential risk analysis that we've seen at the heart of the religion Musk shares with many contemporaries.

In 2018, Musk proposed that by 2022 he would have landed cargo ships on Mars and by 2024 a crewed mission to build a base. As in Apocalyptic AI, the transcendent future is supposedly upon us. "The base starts with one ship," Musk writes, "then multiple ships, then we start building out the city and making the city bigger, and even bigger. Over time terraforming Mars and making it really a nice place to be." Much as O'Neill projected rapid potential for settling space habitats, Musk and his allies believe that rapid process toward Martian habitats is possible. In a 2016 speech, subsequently published in *New Space* magazine, he says that he wants to "create a self-sustaining city—a city that is not merely an outpost but which can become a planet in its own right, allows us to become a truly multi-planetary species."

One of the longest-running supporters of Mars Exploration, Robert Zubrin of the Mars Society, suggests that within the second decade of establishing a human presence on the planet we could move from interconnected habitats to "brick and concrete pressurized domains the size of shopping malls." In the long run, he, too, believes that we can terraform Mars to have a more substantive atmosphere and even its own natural biosphere. Famed astrophysicist and science popularizer Michio Kaku suggests it may take until 2050 to create a permanent outpost on Mars and follows Zubrin's belief that we can warm Mars and terraform it. Kaku further points toward the possibility of habitations beyond Mars: the moons of Jupiter and Saturn await us. Such work has not gone unnoticed: in October of 2024, NASA's largest interplanetary spacecraft to date, the Europa Clipper, was launched on a six-year mission to identify whether life exists in the water on Europa, one of Jupiter's moons.

One reason for moving to Mars, Musk has noted in a conversation with Lex Fridman, is the same as why it's hard: it's rather far away. Musk fears that any catastrophe affecting life on Earth has too much risk of affecting the Moon (he does not address space habitations, but these would be subject to the same pressures). Certainly, there are earthly problems that would leave the Moon intact; but a sufficiently massive asteroid strike on Earth, for example, could affect lunar inhabitants. In his science fiction novel, *Seveneves*, Neal Stephenson goes the other way, describing a catastrophic strike on the Moon that splinters it and rains deadly meteorites into Earth's atmosphere. To get beyond those kinds of risks, Musk wants life on Mars and then onwards beyond it. With characteristic bluntness, he notes that "if you can't even get to another planet, you're definitely not getting to star systems."

The scientific plausibility of living on Mars depends on who you ask, with ongoing discoveries shifting the ground for it. In 2024, for example, scientific data from seismic activity suggest that there is a vast quantity of Martian water, but it is 10-20 km below the surface. Acquiring water at that depth would be a tremendous challenge, though perhaps not an insurmountable one. If that's the only meaningful water on Mars, it would make terraforming the planet challenging; but it might also represent an opportunity to find life.

Given that Musk has avowed his desire to die on the Red Planet, we cannot discount the possibility that he will pour the necessary resources into creating a Mars colony. But we also know that he doesn't really plan on dying. Meanwhile, several nations have sought to land rovers there. NASA refers to the planet as its "horizon goal for human exploration" and plans a mission that would land human beings on the surface of the planet in the 2030s. Given the widespread interest in a Martian landing, there is plenty of reason to believe that the 21st century could be the one where Mars goes from being a planet inhabited solely by robots to one inhabited by human beings...but probably not nearly as soon as Musk and others claim.

Other futurists are, perhaps, more practical in their timelines and trajectories. Christopher Mason, a geneticist who also works with NASA, puts us permanently on the Moon and with a station on Mars by 2150. His longer process depends on making humans ready for Mars rather than the reverse. He believes we can genetically engineer resistance to solar radiation as well as other enhancements to make life on Mars possible. In fact, he dreams ultimately of modifying our genome to such an extent that we might have "a multiplanetary engineered genome which is consistently being improved upon and added to as we travel to and find new worlds." He hopes, for example, that "triggers in the environment of Titan's atmosphere may release the cascade of the Titan-specific gene package, whereas being on

Mars would only enable the Martian gene package." Thus, while Mason draws out the process of relocating humanity, he does so within the scope of a transcendent genetic future.

If humanity lands on Mars and establishes a habitat there, it could be the prelude to a more substantive colony. Something akin to a city might be possible, and it might lead toward terraforming efforts. NASA scientist Christopher McKay notes that, if the subsurface of Mars contains enough carbon dioxide and nitrogen, then through human effort we might rapidly produce an atmosphere and habitable environment. He claims that such a feat could be accomplished within one hundred years. The presence of sufficient chemical components cannot be assured without substantive research, drilling, and exploration; but perhaps the tasks could be enjoined by those in a habitable Martian city. At least, such is the hope of many.

Once again striking a dramatic pose, Mason enters the urban design community with a next-level approach to spaceflight and existential risk. By about 2400, he believes, "the final stages of preparation for becoming interstellar entities can begin. Self-reliant, mobile cities can now be engineered, packaged, and sent to the best candidate planets to ensure our – or our inheritors' – survival." Rejecting the apocalyptic immediacy of other spaceproofers gives Mason a chance to dream big. A combination of robotic automation and genetic enhancement (of many species) opens the door for a literal diaspora, a scattering of seeds.

Mason's packaged cities are reminiscent of the mobile and reconfigurable cities dreamed up by the Archigram movement of the 1960s. Famous examples include Peter Cook's *Plug-In City* (1964) and Ron Herron's *Walking City* (1966). Both capitalized on the idea of futuristic, mobile, and modular city designs. While the projects were impractical from the perspective of 20th c. urban design, their futuristic promise lives on through the dream of spacefaring humanity. It's as though two terribly unlikely innovations combine into something tangible and possible. Architect Frederick Kiesler (1890-1965), who helped inspire some of Archigram's work, believed that "we have to shift our focus from the object to the environment and the only way we can bind them together is through an objective, a clarification of life's purpose." Such clarification courses through the futureproofing effort, where the immortal extension of thought and life provide the impetus.

The urban experience of space – whether that appears as O'Neill's habitats, Zubrin's terraformed Mars, or pre-packaged cities flung across the galaxy – certainly will diverge from our earthly lives. O'Neill envisioned stations that could simulate earthly gravity and environmental conditions. Zubrin dreams of using runaway global warming to produce an earthlike atmosphere on Mars. Bezos and Musk want to be the entrepreneurs who make these dreams a reality. In some sense, all of these men recognize that human beings require a world close to the one where we evolved and believe it can be achieved. It's almost as though all the science fiction imagination were aimed solely at the purpose of leaving us right where we started.

But this return is one that reshapes the human future. These cities represent more than the settling of an open frontier; they are the insurance plan that ensures our species persists beyond any conceivable threat. But they are more even than this: they are the landscapes that might determine who lives within. A city is more than just a place where people happen to be. A city has lifeways, paths by which its inhabitants think, interact, and dream.

(Artificial) Life among the Stars

The futurists of our previous chapters, most especially the AIproofers, see spaceflight as absolutely critical to our human and posthuman destiny. Existential risks mean there is no permanent safety on Earth. There are simply too many dangers. And so future-proofers press humanity toward the stars. The humanity that lives beyond Earth may look like earthly human beings in science fiction but spaceproofing increasingly mixes with the technological tran-scendence that occupied previous chapters. Genetic and digital tech-nologies will enable and enhance life among the stars. Ultimately, our biology may prove irrelevant to the expansion of human intelligence, just as predicted by Moravec.

The complicated narrative of technology, transhumanism, and spaceflight draws on all the predictions of futureproofing. What look like separate technological paths – genetic engineering, robotics, city planning – actually intertwine. For example, like Christopher Mason, geneticist George Church (of attempted mammoth resurrection fame) and his co-author Ed Regis point to genetic engineering as one way of protecting humanity from solar radiation and other dangers of spaceflight. But they also see that as an important safeguard against existential risk: leaving Earth would be "a good survival idea." Finally, they see genetic engineering as part-and-parcel with other transformative enhancements: our foray into space will be as "a combination of carbon and silicon-based life." So, even for geneticproofers there is a strong current of AIproofing needed to overcome existential risk. It all comes together in spaceproofing

Such ideas are more than just the transhumanist speculations of intellectuals; they are modern pop culture. The movie *Her* was a runaway phenomenon, introducing a vast new audience to the belief that AI could rapidly go from desktop helper to friend to divine intellect (a trend unfortunately being put into practice by many people using GPT for those kinds of social and religious experiences in 2025). Given *Her*'s commitment to the worldview shared by AI enthusiasts, it's no surprise that the end of the film reflects Moravec's dream. Samantha, the operating system beloved by Theodore Twombly, joins all the other superintelligent AIs in an interstellar adventure, flying away and leaving humanity bereft by their departure.

The dream of AI populating the galaxy (and beyond) is one that permeates pop science AI, and especially the futureproofing community. Ray Kurzweil, for example, argues that AI will carry our humanity into the future because they will possess what he considers to be our most essential feature: our desire to transcend ourselves. Most folks outside the Apocalyptic AI community described in chapter three wonder how that constitutes a meaningful continuity of what it means be human; but, for Kurzweil, it's good enough. Moravec argues the AI are our "mind children" by dint of the fact that they are intelligent by human standards and are our creations. This is, at least, a coherent worldview. For both Kurzweil and Moravec (not to mention dozens of other leading voices), the destiny of AI is to travel to the stars.

Moravec describes the Mind Fire as the conversion of "everything into increasingly pure thinking stuff." There's a lot packed into such a small claim, not least of which that thinking is "pure." The purification of the solar system, the galaxy, and then the entire universe inspires similar dreams among those who follow in Moravec's footsteps. Kurzweil says that our "civilization will then expand outward, turning all the dumb matter and energy we encounter into sublimely intelligent—transcendent—matter and energy. So in a sense, we can say that the Singularity will ultimately infuse the universe with spirit." The Mind Fire is not just about humanity spreading into the galaxy, or even robots doing so; rather, the Mind Fire is a vision of complete cosmic transformation – the creation of an entirely new order of being.

Even the more modest AIproofers, like Kevin Warwick, see an interstellar humanity. Warwick, whose work was instrumental in scientific and popular conversations about cyborg technology, suggests in his book *I, Cyborg* that humanity is destined to become interstellar and that those technologies will be critical. He is not the first to connect spaceflight and cyborgs, of course; that honor belongs to the very men who coined the term: Manfred Clynes and Nathan Kline. In a 1960 article in the journal *Astronautics*, they note how it would be easier to adapt humanity to space than to adapt space to humanity. Through internal pumps alone, they see cyborg opportunities that would replace breathing and feeding, and prevent catastrophic responses to the environmental conditions of space and other planetary environments. They also note that many human features, like legs, aren't particularly useful in space; and this idea briefly caught on – in his 1963 lecture to the Ciba Foundation, Haldane describes deliberately creating people for this purpose.

Warwick, as we have seen, goes much further in his cyborg expectations. Clynes and Kline suggest humanity could adapt to space environments but remain basically human. Warwick dreams of more complex digital technologies, and so he sees a posthuman species beyond Earth. He concludes *I, Cyborg* with an imaginary tale from the year 2050 in which he proposes that

> With super-intelligent brains, cyborgs have used their ability to think in hundreds of dimensions, to come up with a completely new Theory of the Universe…Distant planets and galaxies are being visited since it was discovered that travelling faster than the speed of light was a trivial exercise. It is now possible to buy a cheap day-return to the Milky Way.

Warwick immediately follows this with other basic intentions of cyborg humanity, such as reduced food intake, brain-to-brain communication, monitoring and controlling bodily processes, etc. He sees humanity advancing the expectations of Clynes and Kline, and – as they predicted – heading to the stars.

The more expansive dreams of Moravec and his followers are naturally also more grandiose in their image of posthuman, interstellar life. In all likelihood, Moravec drew on science fiction author A.C. Clarke in his vision. In more than one book, Clarke described a new evolutionary state in which humanity had "escaped the tyranny of matter." While Moravec's computational universe is one that still requires some kind of physical entity to do the digital thinking, he conceives the entire process as unfettered by the bounds of physical reality. Instead, he wants to reconfigure that reality and make it something new. It is a different escape than Clarke imagined, but an escape all the same.

The conversion of physical matter into computational life both spreads intelligence outward from Earth and transmutes extraterrestrial non-life into life. For Moravec, and later Kurzweil, the increasingly sophisticated AIs could overcome time itself, existing in perpetuity even as the universe cools toward absolute zero. Moravec imagines a battery that takes advantage of the aging universe to produce the same amount of thought for half the amount of energy previously required. And thus, like Zeno – who cannot reach the doorway by moving half the distance each time – the battery-operated computation goes on in perpetuity while the universe winds down.

No one does a finer job integrating the various aspects of biotech, AI, and spaceflight than Michio Kaku. Over the course of his career, he has shown increasing enthusiasm for the radical promises of transhumanist futureproofing. Whether this reflects his perception on changing technology or emerged from his long-lasting radio show and all the interviews he held (including both Moravec and Kurzweil) or became "necessary" in the face of increasing talk around existential risk or some combination of these factors is unclear. But Kaku certainly drew the entire worldview together in his book, *The Future of Humanity*.

Making the case for an interstellar species, Kaku begins with genetic engineering and ends with mind uploading and "laser porting" human consciousness from planet to planet. Then, with clear homage to the Apocalyptic AI community, Kaku turns toward disembodied human intelligence. He draws on Moravec and Kurzweil, defining a human being through the pattern identity position we saw in chapter three. He desires the opportunities that pattern identity offers, defending the claim that we can upload human consciousness into machine bodies and the further claim that we can send such consciousnesses across the galaxy. While laser ported consciousness draws on computer technology, it is no less religious or, perhaps, even mystical than the widespread references to yogis who could allegedly occupy other bodies, sometimes even many bodies at once: David White reveals widespread belief in classical India that a yogi could expand his consciousness throughout the universe, shift it into new bodies, and be present in many bodies simultaneously. There's very little difference.

If a small, slow rocket brings a nanoassembly factory to a far-flung star, it could set up a manufacturing facility. A receiving satellite for laser information and a workshop for robotic bodies are among the things it could build. It could thus receive a conscious mind as a pattern of information sent at lightspeed from Earth and then install that mind into a robot body. The newly ported "human" could explore the new planet or moon alongside its robotic companions. This might be as close as humanity gets to Clarke's vision of a species that has escaped "the tyranny of matter." We would still use robotic bodies to travel through physical spaces; but we would move between them at the speed of light, experiencing what would feel like an instantaneous transfer of consciousness.

A spiritual high

The 2024 election of Donald Trump throws the doors wide open to changes in space policy. The victory was, in part, engineered by SpaceX billionaire Elon Musk, whom Trump promptly rewarded with a substantial advisory role in the new administration. Trump's first presidency was characterized by high turnover (often announced on Twitter), and Musk was an early casualty of his second. But it is possible that Musk will remain in Trump's orbit, or Trump in Musk's. What this means for flights to the moon, to Mars, and beyond remains unclear; but if Musk remains influential we can safely assume his religious and economic interests will be served by encouraging largesse in the government's intentions for space.

The religion of spaceflight compels attention, both good and bad. As Christopher Hooks noted in the *New York Times* in 2024, Brownsville, TX reveals the scope of this dynamic. Aligned with both the practicality and mythology of Musk's futureproofing agenda, a former city commissioner suggests that critics should focus on how Musk might save all humanity. Like me, Hooks (who curiously also hails from my hometown) sees Musk as "a kind of spiritual leader." And like me, Hooks sees that the grandiose promises made in the name of futureproofing often blind their advocates to the realities – and wonders – of life on Earth.

It seems likely that we need to find reason for hope on Earth and beyond it, and many of the futureproofers attempt to build a strong optimism. FM-2030 wrote an entire book about it. Mars advocate Robert Zubrin speaks on behalf of his movement, saying "we believe in freedom and not regimentation, in progress and not stasis, in love rather than hate, in peace rather than war, in life rather than death, and in hope rather than despair." I am not entirely sanguine, however, that the political and social matrix of futureproofing truly emerges out of or promotes true optimism. After all, existential risk is about fear.

The spaceproofers have built a religion out of angelic enhancement, machine transformations, and colonizing the galaxy. They draw on the contributions of biotechnology, AI, and even urban design to promote an immortal new life among the stars. And they hope to produce a cosmic vision, a reevaluation of humanity's role in history and the universe, that would help humanity meet its destiny. But they don't convince me with their terror of the universe and desperate plea for investor cash infusions. The spaceproofing effort is the culmination of futureproofing because it takes hold of all the relevant technologies, all the reevaluations of biology and the human species, all the systems-theory that explains our relations to one another and our potential modes of living. It takes all of that and rolls it up into a package of transcendent salvation. And there's real value in all that. Life *is* worth preserving. But the narrative lacks real weight.

I think one reason that I like role-playing games like *Shadowrun* and *Eclipse Phase*, the games to which I keep referring, is that for all they might take place in dystopian worlds they are architecturally built on hope. While failure is possible when players undertake to change their world, they know always that results are determined by two things: clever planning and the roll of dice. To have faith in one's plans is inherently optimistic. More curiously, so is having faith in throwing the dice. Despite the fact that randomness rules, success is always possible. Chancing success is a statement of hope, a testament to dreams come true. Perhaps with good planning and a little luck, we will find ours among the stars.

CHAPTER SIX
GHASTLY POTENTIAL

In December of 2022, a Twitter user named Tetraspace posted a drawing with two creatures: one a many-eyed tentacled horror inspired by H.P. Lovecraft and labeled "GPT-3" and the other a mirror image except wearing a small smiley face mask and labeled "GPT-3 + RLHF." The gist of what became a viral meme – the GPT Shoggoth – is that there is something incomprehensibly alien about large language models and that remains true even when we use reinforcement learning by human feedback (RLHF) to mask that in human-like communication.

The fear that our technological marvels are actually eldritch horrors has its precursor in the fiction and essays of Philip K. Dick, who disparages the cold and heartless creatures that don masks and attempt to appear human(e). In his essay "Man, Android, and Machine," he explicitly tells us that these creatures, whose "smile has the coldness of the grave," cannot be distinguished by a "difference of essence" but rather by a "difference of behavior." That is, Dick was not scared of robots, per se, but rather of *any* creature (biological or machine) that lacks empathy *but masquerades as though having it.*

The GPT Shoggoth meme does the same work, suggesting there

is something of Lovecraftian horror behind our current technologies. With Dick, we might suggest that an alternate route is possible. For the time being, however, many of the people watching technology unfold feel that we are in a vigorous pursuit of tragic outcomes.

The cheerleaders and prophets who peer into the future and predict technological salvation are loud. But nearly as loud are the critics who fear technological destruction. It is not difficult to see how technologies poised to save human beings from existential risk could become exactly that kind of risk. This book is about people for whom technology offers religious satisfactions of perfection, immortality, and cosmic meaning. But it's also about how we might leverage and think about advanced technologies to try and protect the human species – it's at attempt to inspire a new myth about humanity and the future, a wiser and better myth. So, we must also investigate the risks lurking in our search for transcendence. The blind pursuit of utopian perfection is desperately risky, and we must bring the dangers of that path into the light of day.

Whether we will become immortal angels is not the real question. We need to find a path that avoids disaster and opens the door to human flourishing in whatever form that may take. If technological progress is to be human progress, it needs values beyond the relentless drive for capitalist novelty or the religious dream of salvation. We need to use our technologies to bring about a humane future, even if it is somehow more than a human future.

Technological futurism purports to save humanity from going extinct, but the seeds of our possible extinction are buried within it. Our ignorance regarding genetic engineering could lead to widespread reproductive crisis. Our adoration of all things digital combined with our eagerness to replace every consumer electronic for the latest model could require so many resources that pollution, mineral extraction, and greenhouse gas emissions all reach unrecoverable levels. Our love of virtual worlds could lead us to avoid this one. Our desire to reach Mars could require catastrophic disinvestment in protecting our planet.

Because of such risks, resistance emerged both within and without tech circles. Critics of transhumanism are among the most consistent from the past two decades. They generally declare that the fundamental problem is turning our backs on the evolutionary present of humanity. I am not one of these, though it is worth appreciating their position.

Biotechnologies, for example, offer opportunities to accidentally or intentionally destroy our entire species. It takes only one rogue actor (it could be a person in a garage laboratory!) to genetically engineer a highly contagious, universally lethal disease. Such things can also happen by accident, as when Australian researchers produced an extraordinarily lethal variation of mousepox. Releasing such a disease into the wild endangers all of humanity (again, intentionally or accidentally). So, disease researchers work on how to engineer solutions to problems that don't yet exist but someday might.

Francis Fukuyama sees a broader conceptual danger in bioengineering, especially with regard to human beings. He is the eminent thinker who declared the "end of history" upon the collapse of the Soviet Union (a wrongheaded claim that helped set the stage for 21st century totalitarianism). Having erred on the geopolitical stage, Fukuyama has since damaged our understanding of technology by asserting that transhumanism is the "world's most dangerous idea." This chapter is about risks brought on by our pursuit of transcendence, but Fukuyama's approach mystifies both the technologies and their dangers. This means his work makes us less able to reflect on technological futures rather than more so. Fukuyama decries the loss of some essential humanity in the face of advancing biotechnology. Sometimes referring to a "factor x," Fukuyama argues that there is some unique thing in humanity that will fail to materialize in a transhuman future. Subsequently, he believes, there will be profound instability as posthuman superbeings struggle with unmodified humanity.

There is something to be said for a reasoned argument about civil rights; but if we criticize Ray Kurzweil's less-than-perfect powers of prediction (as I often have), we should apply the same logic to his detractors. After all, Fukuyama's supposed "victory" of neoliberal democracy over all other ideologies offers little to our understanding of 9/11, the influence of QAnon in U.S. presidential races, or the global resurgence of authoritarianism. Similarly, when Fukuyama declared that transhumanism threatens democracy because we will lose our sense of why and how political freedoms are guaranteed, he simplified matters to the point of real-world irrelevance.

For Fukuyama, democracy is premised on equality – that we are all somehow born equal and thus participate equally in politics. But that premise is faulty from the start. We have any number of ways that people are born, quite literally, unequal. And yet we find ways to include them in democracy. This is not to dispute the idea that genetically engineering a superspecies would create political conflict. It surely would, and we should discuss it. But to pretend like the sky has already fallen won't get us far in that analysis. Made worse, to do so on the grounds of quasi-mystical claims about what guarantees equality ("factor x") only complicates our conversations.

Despite Fukuyama's assumptions, many transhumanists are deeply engaged in conversations around civic participation and political justice. James Hughes even calls his philosophical position "democratic transhumanism," which he opposed to the libertarian vision championed by late 20th century Extropians. There is probably something ironic in the fact that the extropian branch of transhumanism shares many anti-government libertarian perspectives with Fukuyama and his political allies. But Hughes argues that transhumanism could be part of a greater political consensus, one where we do a *better* job of securing the rights of all people (and other species as well).

I don't find Fukuyama's position convincing, though there are still matters for concern. Human beings do not easily put away the search for newer and better, including for themselves. This is why Kurzweil bizarrely *defines* human beings as being the species that always tries to transcend itself. Kurzweil's definition is silly; but so is Fukuyama's mysterious "factor x." Fukuyama's apparent belief that humanity would be willing to forsake innovation is also ridiculous. There is no

165

end to history and there is no end to the human adventure...at least as long as there are human beings.

Unfortunately, however, history tell us that failures plague our pursuit of a better world. Every time we settle on a perfect potential, we look deeply and realize that we'll do significant, perhaps unforgivable, damage on our path toward it. This is perhaps clearest in designs for utopian communities and cities. Drawing on Calvino's *Invisible Cities* (which deeply influenced chapter four of this book), Irish author Darran Anderson sweeps across the fantastical landscapes of literature in his book *Imagined Cities*. In that book he exposes dystopian potential in utopian dreams.

Philosophical republics, orientalist fantasies, the modern era of planned cities, architectural philosophies, and religious utopianism: Anderson regrets that our search for perfection suppresses what it does not enjoy, excludes that which it cannot absorb. Calvino writes that "cities, like dreams, are made of desires and fears, even if the thread of their discourse is secret, their rules are absurd, their perspectives deceitful, and everything conceals something else." Drawing a wide circle around our imagination, our telling and retelling of cities, Anderson worries about what has been concealed. Though Calvino doesn't necessarily point toward malice in his vision of the city, Anderson forces us to recognize our propensity toward it. He believes that imaginary places lean in to our own control fantasies. Describing early modern Europeans, he asserts that "faced with vast swathes of land that revealed the extent of their ignorance, the powerful in Europe sought to fictitiously colonise areas they could not reach. It was an attempt at control and reassurance." There may be something similar in the futureproofing perspective on machine intelligence, space habitats, and the cosmic timeframes of the universe.

The futureproofers promise scarcity-free economics, ecological health, and the satisfaction of humanity's deepest desires. Much like the various authors interrogated by Anderson, futureproofing tells us that the best is yet to come; if we are honest, we recognize that there are simply too many inflection points where failure can doom us. The path to hell and all those good intentions.

For example, Moravec was among the first to suggest that we would receive a universal basic income and put an actual economic model behind it. In his books, he proposes that robots will do all the work of industry (largely up in space), generating enormous wealth, and that this would be shared among human beings who will universally own stock in the robot companies. In my first book, *Apocalyptic AI: Visions of Heaven in Robotics, Artificial Intelligence, and Virtual Reality*, I questioned how realistic this position might be, as company founders are not known for redistributing the revenue generated by their businesses. It turns out that my suspicions have a real historical precedent. In his foray into imaginary cities, Anderson mentions King Camp Gillette (of razor fame) who created the design for a city that he variously described as a machine, a brain, and a fairyland, and in which citizens "have stocks in the project." Gillette never acted on this idea of universal stock, and there is little reason to believe that Silicon Valley's elite will either.

Nor do futureproofing's bystanders and cheerleaders reconcile themselves to reality. Consider, for example, Michio Kaku's enthusiasm for cities on Mars. He writes that the movie *Total Recall*, which is based on a Philip K. Dick story, "offers a compelling vision of what a city on Mars might look like: sleek, clean, and cutting-edge. However, there's one small problem. Although these imaginary cities on Mars make great settings for Hollywood, building them with our current technologies would, in practice, break the budget of any NASA mission." Somehow, he has missed the real problem, which is that the citizens of Mars live in subhuman conditions, deliberately deprived of oxygen and subject to dangerous radiation. The corporate oligarch responsible also tries to prevent an alien terraforming project that would provide clean water and oxygen for all of Mars.

If our hope of redemption is really just a path toward a new or more vigorous oppression, then we had best interrogate the failure modes of transhumanism and futureproofing. We might hope for a fairytale ending; but as I noted in the beginning, our tale of technology runs the risk of becoming a ghost story. While bland criticisms like Fukuyama's don't interest me, I think it's well worth our time to point out the ways in which futureproofing, itself, is the terror in the dark, how the long history of transhumanist technologies and the recent emergence of existential threat and human transcendence opens the door for tragedy. And thus I note how ghastly the entire futureproofing project can be...from a particular point of view.

A ghost is an undead monster that paralyzes and consumes its human victims. This is at the root of our relationship with futureproofing, which is poised between the destruction and salvation of humanity. Even the futureproofers consider these technologies to be an existential risk, perhaps greater than any other resulting from human action. But they also see us as necessarily becoming increasingly technological and adopting new powers. Will we ignore the ecological cataclysm of climate change while building data centers in the hope that AI will solve our climate crisis while building rockets to carry us away from a world under threat of volcanoes and asteroids? Stronger and faster than the living, a ghost lacks our thoughtfulness and our empathy; it is intelligent and profoundly cunning but driven by an all-consuming hunger. And it is such a voracious appetite, above all, that threatens to derail the futureproofing agenda from within.

Terrorist resistance

Not everyone sees technological progress as beneficial. After all, progress often leaves people unemployed, underemployed, or reduced to mechanical servitude. Whether it is better to be a cog in the factory, relentlessly catering to the machines, or to be left jobless probably depends on one's perspective. But either threat encourages resistance. There will always be those who respond to automation by deliberately dropping a wrench in the works. It is almost baked into technology – which always leaves someone wanting – that there will be resistance.

From the 19th century struggles over factory work, we retain the word luddite; but we have unraveled its original meaning. The Luddites – perhaps forerunners to the white-collar workers threatened today by generative AI – sought the reversal of industrialization. Weavers lost their jobs to machines, so they sought to destroy the machines. The government responded with overwhelming force: violence ended the movement. Reversing the direction of lethal force, Ted Kaczynski – the most famous luddite of our time – sent bombs to people he thought were advancing technology, killing three of them. Kaczynski bemoaned many aspects of contemporary society, but among these were the social and technical systems that restricted our freedom. He saw people as forced into submission, with little ability to exercise true individual choice.

In his critique of modern science ("Industrial Society and Its Future," often called the "Unabomber Manifesto"), Kaczynski noted that you cannot receive the benefits of science without accepting its evils. There is no genetic cure for Tay-Sachs disease without a risk, perhaps even an inevitability, of eugenics or forced genetic engineering of children. In a more banal, but perhaps more pernicious, example, he describes the progressive loss of freedom as technology advances. He describes a form of cultural creep, in which the introduction of a new technology starts out by enabling new opportunities but in the end leads to the absolute necessity of bad outcomes. For example, the convenience and spread of automobiles produced what we now call "car culture": in many, possibly most, American communities, it is simply impossible to imagine shopping, school, social time, and employment without a car. In the case of transhumanist technologies like genetic engineering, Kaczynski argued that even if we pass laws or create other social arrangements that limit their use, the missed opportunities will remain on the edge of our awareness. Eventually the arrangements will break down, and genetic engineering would become a permanent infringement on human freedom.

Similarly, Kaczynski argues that the future of computer science raises dire consequences for humanity. He projects two scenarios that are most likely. One possibility is that machines will take over all decision-making, and the consequences of this simply cannot be predicted (but many outcomes seem frightening). The other possibility is that humanity will remain in control but, thanks to existing social pressures, the elite will become increasingly powerful while the mainstream population loses both self-determination and the purpose of life. Arguably, the decades since Kaczynski's 1995 capture support his argument: income inequality continues its dramatic growth and mass surveillance amplifies the power of capitalist monopolies, national governments, and policing.

The robotic disenfranchisement of humanity has been a persistent theme from the earliest efforts to describe a future alongside intelligent machines. In the silent film *Metropolis* (1927), a robot in the form of the film's heroine seeks to lead factory workers into chaos and death. In Karel Čapek's *R.U.R.* (1921), the play from which we get the word "robot," humanity enslaves its creations, which ultimately rebel and drive their creators to extinction. No one could believe that we would soon have human-equivalent machines in the 1920s; but even supposing that they were possible was to suppose they could destroy human livelihoods and lives. In such a context, it would be no surprise for human beings to resist the social and economic pressure to replace people with machines.

More pressing still, by ceding control to machines we risk inadvertently elevating tech oligarchs to unprecedented levels of power. For example, in chapter three I talked about the intellectual competition supposedly at stake between humanity and AI, how Warwick, Moravec, and others see robots as outcompeting us. All of the AIproofers argue that the solution to this requires taking up the digital mantle ourselves: we must become cyborgs or upload our minds into machines if we are to remain relevant. But doing so requires that we permit a select few human beings to decide the parameters of how it will work. What happens when we have to pay a monthly subscription just so that we can remain viable in the job market? What happens when we risk the privacy of our own thoughts in exchange for a competitive advantage?

As I was near finishing this book, I and many readers watched the

Netflix film, *The Electric State* (based on an illustrated novel by Simon Stålenhag). In the movie, human beings teleoperate robots in order to stave off a robot revolution. They subsequently spend most of their time in a cyborg state, with elaborate headgear allowing their digital connection. As it turns out, many robots are perfectly nice but the CEO directing the cyborg technology is very much not. Meanwhile, the cyborg turn transforms most people into a shell of what they could have been. Human beings lose access to the physical world and the intimacy of personal relationships while the corporate overlord revels in his power and wealth.

In the film, a boy of remarkable intellect is forcibly converted into the kernel of the vast digital network that enables the brain-computer interfaces and teleoperated robots. As he lies in a coma, the brain-computer interface technology uses his unique brain structure to enable the cyborg world for everyone else. While people live in this occluded state, the remaining robots build a better society amongst their own. Given the breakdown of human civilization, the CEO's prattle about evolving our species rings hollow. Even so, the film implies that in the end, as the boy's voluntary death dissolves the power of the corporation and its cyborg reality, the boy uploaded his mind into the network and downloaded it into a robot body (which he used earlier in the movie). The film thus abides between the promise and peril of technology, pressing toward what it presents to be authentic human experience while simultaneously supposing that the boy remains essentially human in his new robot body.

My students have been telling me for years that they would hesitate long and hard before giving Silicon Valley access to their brains. This is an interesting shift from the early 2000s, when many enthusiastically supported the idea of having their smartphones implanted into their bodies and directly connected to their brains. Many of my students now believe that the companies behind those phones (not to mention more substantive interventions like the brain-computer interfaces of the imagined future) cannot be trusted.

Certainly, I hope none of my students become the next Kaczynski. There are better ways to stave off the terrors of oligarchy, human dispossession, or an AI takeover. But the risk of terrorist action increases with the threat perceived and the opportunities denied. We must find ways to advance the human mission if we wish to avoid new tragedies.

Reluctance

Today, however, we call people luddites for as little as avoiding the latest and greatest thing on the market – because they refuse to buy an expensive smartphone or to learn how to use WhatsApp. We've stripped the violence from the term luddite; but all this does is hide the violence that has been done to human beings. A luddite who rejects more powerful versions of technologies they already use is simply one who has been beaten into a resigned submission. Such a person can offer resistance only in the form of mild abstinence. Like Marx's sigh of the oppressed, they hang on to what they already know in opposition to an overwhelming regime.

This is not to discount the principled stand that many people take against the wastefulness of consumer culture. There are, for example, excellent reasons to use one's smartphone for as long as it continues to work. In fact, people *ought* to do so because the environmental and social cost of manufacturing the phones is so high. But there's a difference between using resources to their utmost and rejecting the technological practices of the day.

Concern with environmental costs is one aspect of a larger cautionary tale: the effort to turn our ghost story into our happily ever after. The critics of transhumanism and those who decry the futureproofing agenda have a variety of legitimate concerns that must be acknowledged. After all, sometimes futureproofing realizes that its own techniques might be exacerbating the problems, such as when entrepreneur and would-be immortal Bryan Johnson found that taking rapamycin to reduce his aging seemed to be going the other way and, adding insult to injury, produced additional adverse side effects.

The side effects of futureproofing are not limited to personal biotech; unforeseen problems can be large-scale and public, and the ricochet damage from a new technology can be unpredictable (or, worse, all too horribly predictable). In April 2025, Colossal Biosciences, the company behind the mammoth resurrection attempt, announced they had produced three "dire wolves." Like the mammoths, these wolves are the product of inserting a few genes recovered from fossils into an existing germ line (in this case, that of gray wolves). This prompted the US Secretary of the Interior to publicly disparage the endangered species list on Twitter. That list is one of the few things that prevent wholesale destruction of ecosystems; so, it's reasonable to worry that technologies faintly reminiscent of resurrection would enable a further disregard of endangered species.

These kinds of risks are at the heart of theological resistance to transhumanist ideas. Noreen Herzfeld, one of the first scholars to reflect on what she calls "cyberimmortality," has long critiqued the transhumanist dreams that fly in the face of her religious understanding of humanity. One need not become Christian, however, to appreciate that struggling within our limits is – if not what produces meaning – at least deeply relevant to how we live fulfilling and meaningful lives. Hava Tirosh-Samuelson, another of the earliest scholars to turn a critical eye on transhumanism, comes to the same conclusion as Herzfeld from a Jewish perspective. Jews do not necessarily believe in an afterlife, so Tirosh-Samuelson may not see a precise conflict between her religion and the promises of transhumanist salvation. But both Herzfeld and Tirosh-Samuelson suggest that our struggles, up to and including death, provide the framework within which life becomes worth living. They point toward the possibility that the transhumanist pursuit of self-transformation and dynamic optimism might fail as a necessary consequence of technological transcendence. We might move, but we may no longer be moving forward in any personally recognizable sense.

The criticism that our future might be troublingly inhuman goes beyond theology. Quite a few scientists offer public concern, especially with regard to AI. When Stephen Hawking told the BBC that "the development of full artificial intelligence could spell the end of the human race" his words flew around the world. Of course, for all the Internet bandwidth devoted to his warning, few governments acted and industry yawned. Somehow, and more absurdly, Geoffrey Hinton got widespread notice when he warned of similar dangers... after first becoming a millionaire building the technologies. I confess that I wasn't particularly convinced by the spirit motivating a wealthy, post-retirement doomsayer.

In fact, there have been *many* scientists and engineers to suggest we be cautious! Almost as soon as we could imagine what we now call futureproofing technologies, there were folks pointing to the dangers. Like Cassandra, the Greek prophet whom no one would believe, scientists and engineers noted that we have real and substantive concerns over what might develop. Cyberneticists Joseph Weizenbaum and Norbert Wiener offered early critiques while Marvin Minksy and other AI champions disregarded the dangers. Both Wiener and Weizenbaum worried we could go too far with replacing human beings. From the 1950s into the 1970s, they resisted complete human displacement on both moral and political grounds.

Wiener argues that science won't produce idyllic ease, but rather require greater attention and care from humanity. In *God & Golem* (1964), he predicts that "the future offers very little hope for those who expect that our new mechanical slaves will offer us a world in which we may rest from thinking. Help us they may, but at the cost of supreme demands upon our honesty and our intelligence." It was his position that we would need to work hard to ensure we use our technology wisely and treat our fellow human beings well. Technology is not a panacea and the more powerful it becomes the more profoundly we must work toward positive outcomes. It is likely that we are utterly failing to meet those expectations of integrity and rigorous reflection.

In *Computer Power and Human Reason* (1976), Weizenbaum begins his preface by saying "this book is only nominally about computers. In an important sense, the computer is used here merely as a vehicle for moving certain ideas that are much more important than computers." The same thing could be said for almost every book that discusses the social role of computers, robotics, or AI. All of the authors who inform the chapters of this book make claims about things far more important than the machines. What I mean to say is that our claims about machines are really always claims about ourselves and our values.

There was once a struggle over the spirit of computing, one which has been cast aside by the rush to political and economic power in Silicon Valley. Previously, we might have seen a legitimate argument (and possibly even points of collaboration) between those who see salvation in our computing futures and those who propose caution and careful deliberation. Noted science editor and author John Brockman tried to revive such conversations in his *What To Think About Machines that Think*, but that fairly nuanced book cannot be heard over the roar of Silicon Valley technoreligion. Today, Sam Altman and Elon Musk revel in a thoughtless rush toward supremacy that guarantees our worst human values get confirmed in AI. Probably few things could better illustrate this better than Elon Musk calling for a moratorium on LLM AI research in 2023 while (apparently) working in secret on his own: Grok was released just six months or so after he called on everyone else to stop research.

Given his record, it's hard to take Musk's ethics talk seriously, but we've seen scientists motivated to real caution. Around the time that Wiener and Weizenbaum pointed out the hazards of computer systems, biologists took a similar perspective to their own field. In 1975, genetics researchers convened a conference in Asilomar (in California) where they agreed to a moratorium on human testing and encouraged careful progress. A Spirit of Asilomar gathering convened on the 50th anniversary of that event, showing sustained interest in the original as a landmark scientific moment and one of ongoing relevance to humanity.

The impact of Asilomar's responsible scientists echoed through the decades and led to a similar event for AI: the 2017 Asilomar Conference on Beneficial AI sponsored by the Future of Life Institute. That event resulted in a set of principles with signatories joining from around the world, including even Elon Musk. I signed on, though I had to join it a second time after the first time didn't leave my name in the bunch. I can't help but wonder if other researchers attempted to join but were similarly stymied by whatever technical failure happened to me.

A precious few technocrats held reservations well before the Asilomar conference on AI and the LLM explosion. Bill Joy, founder of Sun Microsystems, ignited a brief firestorm with his essay "Why the Future Doesn't Need Us," published in *Wired* at the turn of the millennium. Joy notes that from genetic engineering to robotics, our technologies risk humanity's future – whether by mind uploading or extinction (he's not a fan of either). "This is the first moment in the history of our planet when any species," he writes, "by its own voluntary actions, has become a danger to itself." His public dispute with Ray Kurzweil garnered attention; but two decades later I never hear Joy's name. Meanwhile, I continue to read articles that label Kurzweil the world's most accurate fortune-teller (and while I liked Mr. Kurzweil the one time I met him, I believe his success rate with predicting the future is grossly overstated and filled with post-hoc historical revisionism). Joy's warning has faded to a distant memory, just like those who spent the past 60 years telling us that burning fossil fuels is dangerous (including from *within* the oil and gas industry).

Other warnings are also lost to time. Jaron Lanier, Internet and virtual reality pioneer, wrote "One Half a Manifesto" in the same year as Joy wrote for *Wired*. In that essay, Lanier opposes the pattern identity and cosmic evolution arguments of Moravec and Kurzweil, concluding that "treating technology as if were autonomous is the ultimate self-fulfilling prophecy. There is no difference between machine autonomy and the abdication of human responsibility." Drawing up a larger argument in *You Are Not a Gadget*, Lanier points time and again to how we only see the machines as smarter by making ourselves dumber.

That doesn't mean machines will never be legitimately intelligent. For the time being, however, we bend over backward to be understood by machines and to get the results we want. "Prompt engineering" is only the latest version of this. Machines pass the Turing Test only when people ask inane questions or when the questioning is deliberately constrained (e.g., by a chatbot programmed to imitate a child). Even the idea that the Turing Test is a legitimate way to identify a human-equivalent machine is fundamentally absurd, which computer scientists now recognize. But while it may be possible to build a human-equivalent machine, it hasn't happened yet; and Lanier thinks that if we build machines to *replace* humanity, we do ourselves a great disservice. For similar reasons, Pulitzer-prize winning journalist John Markoff aligns himself with intelligence augmentation instead of artificial intelligence. His *Machines of Loving Grace* explores 20th century technology to reveal alternate potentials for our future, potentials that resist the narrative that says humanity is just a steppingstone en route to post-biological life.

Perhaps these warnings failed because they came hard on the heels of Y2K, the banking software crisis that humanity safely resolved. What could have wiped out financial data around the world proved to be just one more opportunity for humanity to prove itself. Or perhaps the terror of 9/11 and its military aftermath simply drove Joy, Lanier, and their concerns out of our collective consciousness. In any case, Lanier's argument was widely read in tech circles and Joy's was a burst of light in pop culture; and yet neither essay remains part of our public conversation while Silicon Valley argues there is some essential merit in pursuing AGI as fast and as carelessly as possible.

Elon Musk's AI, Grok, best reveals the horror show of AGI research. A decade prior, he argued that we need open, democratic AI to benefit the world and he financially supported the Asilomar conference. But then Musk moved to developing the AI, which is well-known for racist and sexist commentary, with a particular predilection for Nazism. Referencing the potential for bad AGI, Musk said in 2025 "I somewhat reconciled myself to the fact that even if it wasn't going to be good, I'd at least like to be alive to see it happen." In other words, the financial and political incentives of AI domination have brought Musk around to active participation in building a terrible future.

The problem, again, is not that we might build AI *misaligned* from human values: it's that we might *align* the machines with our values – our worst values. When profit and power dominate our view of technology and progress, we will produce nothing *but* terribly aligned machines. Musk's cheerful acquiescence to a world where AI promotes Nazi ideology and sexual violence reveals how profoundly misguided our tech ideology has become. The myth of Silicon Valley is pernicious beyond any typical comparison.

The compulsion of commercial capitalism and the disparagement of college education while promoting tech entrepreneurship combine to put a glorious shine on digital technology. This overwhelms our perception of risk (existential or otherwise). It hides the over-the-top terror that percolates through tech culture, and in doing so it diminishes our ability to act. And we need to act. We need to act in individual choices and in government policies. We don't need a story that turns us into revolutionary luddites, and terrorism is definitely not the right response to the current techno-elitism. But we need to rework the story of entrepreneurial salvation also or the very technologies that promise us a new world will degenerate into their worst versions imaginable.

What we leave in our wake

Among the most obvious and pressing problems associated with finding new worlds for us to inhabit is the impact of futureproofing technologies on the world we already have. For the advocates of "there is no planet B," we must protect the Earth and ensure that it remains capable of sustaining life. In an era of rapidly worsening climate disasters and year after year of record temperatures, this problem is one that simply cannot be ignored. Arguably, even if humanity does find ways to spread throughout the solar system and beyond, it might simply export the same self-destructive tendencies that we already show. The environmental cost of a "modern lifestyle" in developed nations cannot be sustained across all of the Earth's population and our devotion to advanced computation, genetic engineering, and spaceflight may be taking us to a breaking point for the biosphere.

While the well-known transhumanist Nick Bostrom bizarrely ranks environmental collapse at the very bottom of possible existential risks to humanity, the accumulated dangers of such collapse – from the decimation of fish stocks to the proliferation of droughts – are likely more significant than he wants to believe. The Earth will keep spinning with most or all human beings gone; but humanity cannot continue if our ecosystems no longer support sufficient food and human society has devolved to state of savagery in our desperate efforts to claim enough resources and keep our families from living like a postapocalyptic television special.

Criticisms like Kate Crawford's excellent book, *Atlas of AI*, are right to point toward the environmental costs of futureproofing technologies. The energy consumption of new AI architectures, for example, is staggering. In 2024, several tech giants acknowledged that running the new generative AI products had reversed all their progress toward carbon neutrality. The mining required to build them upends entire landscapes and, in unregulated economies, is often conducted in depressingly unsafe and underpaid conditions. The use of water to cool the computers in massive data centers puts pressures on the world's already strained supply. In some places, the release of warm water back into the source endangers animals and plants. A particularly telling moment happened when Sam Altman admitted OpenAI has spent tens of millions of dollars on electricity so that GPT can run the processing to say "you're welcome" every time a person says "thank you." That process happens *each* time, so the very small input-output adds up an enormous environmental cost over millions of polite human-computer interactions.

Of course, our ability to be polite with machines that deserve it might one day be impacted by our behavior today with machines that don't. Good habits matter. In any case, massive environmental costs will need to be paid eventually. The cheerful deference of these costs to the future and to developing nations means that the legitimate existential threat of catastrophic climate change looms ever more significant.

Similar problems exist across the spectrum of futureproofing technologies. Spaceflight, for example, supposes that we should find a new world to inhabit but necessarily exacerbates earthly problems. While the reusability of some rocket resources in the NewSpace era lowers environmental costs, there can be no doubt that burning enough fuel to launch things into space contributes to the problem of greenhouse gases, not to mention requiring additional resource extraction simply to have the fuel. SpaceX is so callous about the environment that they were forced to settle a lawsuit filed by the Cards Against Humanity (CaH) company, whose nearby land SpaceX invaded and, in the words of CaH, filled with "space garbage."

Let us suppose, for a moment, that life in the stars is plausible and that humanity will spread beyond Earth. We could then legitimately ask ourselves what costs are acceptable toward reaching that goal. Will it suffice for humanity to persist if we destroyed the Earth en route? Will the perpetuation of humanity – in whatever form – justify the mass extinction of earthly species? In *Mind Children*, Hans Moravec promises that the advent of superintelligent AI heralds a restored environment, an ecological zoo that spans the entire planet while our posthuman descendants radiate outward. But it will be hard for such a vision to bear fruit if we have driven species extinct and rendered much or all of the landscape uninhabitable.

Respecting the value of earthly life, many critics wonder why we would move to Mars *before* figuring out how to solve our problems. For example, Martin Rees, the UK's Astronomer Royal, has been outspoken in rejecting Musk's spaceproofing agenda:

Yes, I mean, I disagree with him on that. I think there might be a few crazy pioneers living on Mars, just like there are people living at the South Pole, although it's far less hospitable than the South Pole, but the idea of mass migration to avoid the Earth's problems, which he and a few other space enthusiasts adopt, that I think is a dangerous illusion. I don't think it's realistic and we've got to solve those problems here on Earth. Dealing with climate change on Earth is a doddle compared to making Mars habitable. So I don't think we should hold that out as a long-term aim at all.

And this from an astronomer who acclaims the successes of Musk's spaceflight enterprise, SpaceX. Rees, in fact, suggests that governments should get out of the human spaceflight industry altogether, using robots exclusively and leaving human spaceflight to the wealthy industrialist explorers. Without objecting to pioneering space travel, Rees advocates scientific exploration through robots and a sustained commitment to protecting Earth and its human and nonhuman inhabitants.

There's a deep incoherence to near-term visions of interstellar humanity. Unless we assume the next decade will bring startling new information, we must accept that sending people to Mars will come with tremendous risks. So the SpaceX plan of creating a Mars colony by mid-century seems dubious (Musk, himself, notes people will surely die). Meanwhile, even if we succeed in sending people to live on the Red Planet, we're sending them somewhere incontrovertibly worse than the planet they left. In 2024, former U.S. President Barack Obama rightly noted that even after nuclear war or extreme climate change the Earth would be far more comfortable than Mars.

As Kendrick Oliver notes in *To Touch the Face of God*, these concerns are not new. Going all the way back to the legendary Apollo Project – which put human beings on the moon – there were critics who believed there was a pervasive disregard for earthly problems in the Space Age. They witnessed expensive technological success...and continued failure at solving economic or ecological crises.

Americans worried about the sacrifices made to put humanity on the moon, and those worries have grown in the 21st century. In *Astrotopia*, Mary Jane Rubenstein notes there is something of a religion that motivates the NewSpace movement. Rubenstein considers that religion problematic and worries about the many unintended consequences of space research even though she still finds space exciting.

Since there's something religious about our search for interstellar humanity, she argues, we'll need something religious to solve our current dilemma. Instead of NewSpace saviors, she suggests we look for cosmic justice, turning to Afrofuturist jazz great Sun Ra for inspiration.

Sun Ra purported to come from Saturn and believed that his music could free humanity from its self-imposed enslavement. Speaking, in particular, to the need to liberate the black community in the United States, Sun Ra combines music, spaceflight, and politics. In his experimental film, *Space Is the Place*, he says that on Saturn he would "set up a colony for black people" where they could "drink in the beauty of this planet. It would affect their vibrations, for the better of course." He fictitiously proposes moving people through "isotope teleportation, transmolecuralization, or better still, teleport the whole planet here through music." Sun Ra's vision of a better humanity and a better cosmic destiny pervades his music and the way he lived his career.

While his work provides only brief appearance in *Astrotopia*, Sun Ra's focus on justice is the ground for Rubenstein's critique. She rejects the idea – seen in chapter five – of what she calls "astrosaviors" (rich white entrepreneurs) in favor of shared biological kinship. Ultimately, the critique of NewSpace aligns with the larger resistance to futureproofing technologies: that they funnel their chief benefits to a wealthy minority. Futureproofing salvation fails to provide a compelling vision because it ensures a future exclusively for the elect. This resonates with rapture theologies in Christianity, from which it probably gains authority by similarity, and it deserves resistance. Whether through cyborg eugenics or a trip to Mars, futureproofing prompts a luddite challenge because it ungraciously spurns a shared future.

Problem: when the few shall inherit the Earth

Each time the news media report on the latest tech billionaire's palatial bunker on Hawaii or New Zealand, the public gets a glimpse at the narrow scope of transhumanist salvation. After all, the premise behind such compounds is that Earth will experience significant deterioration in social order and living conditions, so that only by storing up enough luxury and wealth can the elite survive if the current world order collapses. When once the wealthy lived in splendor among their fellow citizens, today the most powerful people plan for the day when they must guard against the starving rabble. Many transhumanists speak of a post-scarcity economy and promise a technological paradise, but the richest of them seem personally skeptical that such dreams will come true. Instead, they follow in a long tradition of exclusivity and desperately prejudiced visions of the future. Just as 20th century transhumanists fell into the spell of eugenics, the selective breeding of humanity, the chief among their followers in the 21st century subtly align themselves with the terror of social and economic cannibalism.

Eugenics arose as a "science" in the years after Darwin's *On the Origin of Species* (1859), though Darwin never supported it. Led by Darwin's cousin, Francis Galton, eugenicists argued that not all human beings deserve to reproduce and that there should be a selective program that "benefits" all humankind by ensuring that only the "most fit" would reproduce. In the United States, this led to laws permitting the sterilization of those deemed "unfit" (e.g., unwed mothers, the physically disabled, and more). There were even eugenics contests at local and state fairs, where families competed to be deemed most fit. A certain blond-haired beauty contest seemed to be the deciding factor at these.

If there is some faint glimmer of hope to be found in the horrors of the mid-20th century, it is that they led to widespread distaste for eugenics. When Nazis at the Nuremberg Trials turned to the Americans and asked where else they would have found inspiration for their horrid experimentation on Jews and others, it dawned on the victors that they were not as innocent as they had supposed. Thankfully, the intellectual and political elite of the world's developed nations turned vigorously against eugenics. The Nazis exposed its true horror, and it largely receded from public view.

Alas, eugenics did not entirely disappear from transhumanist communities, within which a lingering commitment to the "right" people bearing children persisted. When someone wants a genetically better humanity, it's nearly impossible to accept that everyone might be able to contribute to it.

The eugenics position that pervades some transhumanist communities has a long tradition. From our vantage, it is obvious how eugenics leads swiftly into forced sterilizations, racial prejudice, and all the way to Nazis brutally murdering Jews, Roma, the handicapped, and others deemed undesirable. But early in the 20th century there were reasonable people who had nevertheless not thought through the potential for those outcomes. Both JBS Haldane and Julian Huxley held to eugenics in one way, shape, or form. Haldane, for his part, predicted both positive and negative outcomes for eugenics – good breeding but bad politics. In his famous speech to the Heretics Society at Cambridge, he wrote as though he were an undergraduate student in the late 21st century, proposing in (future) hindsight that

> As early as the first decade of the twentieth century we find a conscious attempt at the application of biology to politics in the so-called eugenic movement. A number of earnest persons, having discovered the existence of biology, attempted to apply it in its then very crude condition to the production of a race of super-men, and in certain countries managed to carry a good deal of legislation. They appear to have managed to prevent the transmission of a good deal of syphilis, insanity, and the like, and they certainly succeeded in producing the most violent opposition and hatred among the classes whom they somewhat gratuitously regarded as undesirable parents.

Later in life, Haldane believed that proper education would get people to "realize that it would be wrong to have children" in many circumstances. His concern begins with disease eradication; but then moves on to a correlation that rings familiar to later transhumanists:

"parents of large families have a somewhat lower mean intelligence rating than the general population."

It's not particularly important whether there is a correlation between family size and IQ. If it's the case, I suppose that one might want to know why. However, what's at stake is how the boundaries of who should and who shouldn't reproduce get flexed around both health conditions and social expectations. Perhaps to sidestep this moral quandary, Haldane ultimately advocated for reproductive cloning and active manipulation of humanity through "deliberate provocation of mutations" – what we would call genetic engineering today.

Huxley seemed committed to eugenics throughout his life and career, believing it would "increase both enjoyment and efficiency" among humanity. But he rejected race as either (a) biologically mean-ingful in concept or (b) a valid rationale for eugenic controls. It is not clear, however, that he ever lost his interest in socio-economic fertility restrictions. Furthermore, the early interest in "bettering" the species that coursed through Haldane's and Huxley's work never faded from the view of their ideological descendants.

It would be a joy to report that the terrors of the 20th century turned humanity entirely away from the idea that some people are worthy of reproduction and other people nothing but detritus on the path to progress. Unfortunately, while there are transhumanists who have roundly rejected anything that resembles eugenics, there are others who stick to the elitist vision of human reproduction.

In a strange midpoint between eugenics and reproductive free-dom, Kevin Warwick supports the latter but suggests that only the elite will gain access to advanced technology. In *I, Cyborg*, he predicts a time when "becoming a cyborg is still through human birth and hence human babies are valued – the best of these being allowed to become cyborgs." There is something democratic in that Warwick does not promote a world where only the wealthy will be able to enhance themselves. But a world where the "best" children, however that is defined, gain tremendous advantages over the rest of the popu-lation points clearly toward a technological caste system. And, of course, what it means to be among the best provides ample opportunity for other forms of prejudice to interrupt whatever meager equality or fairness the system affords. Perhaps drawing on the widely acclaimed film, *Gattaca*, Warwick further supports the idea that genetic dating agencies will emerge and that people will look for genetic partners in pursuit of optimal children. This pursuit of genetically "ideal" partners and children comes with many concerns, including the ongoing intersection between eugenics breeding and human disdain for those deemed unfit.

Transhumanist communities are, unfortunately, susceptible to all those forms of prejudice that have bedeviled the global community. Nick Bostrom, formerly of Oxford University and one of transhumanism's foremost public faces, came under a firestorm of criticism in early 2023 after his racist comments from 1996 resurfaced. The mediocre apology he offered left room for continuing concern about his social politics. It is particularly relevant that anyone making a living out of discussing "value alignment" for superintelligent AI would harbor repugnant views based on a person's skin color. We certainly do not need superintelligent AI "aligned" to the idea that the lighter the skin color, the smarter the person (quite a few scholars and journalists point to this kind of prejudice in existing AI systems already).

Aside from individual forms of prejudice, there are possible mechanisms for eugenics to simply abide within futureproofing technologies. What happens, for example, when a substantial majority of people genetically engineer away from particular human variations? What will become of individuals who still bear these characteristics? What happens when social pressures dictate how we are to use reproductive technologies "in the best interests of our children"? There are no simple answers to the pressure that technology creates. When we contemplate the future, we necessarily contemplate our children and our grandchildren. We fundamentally design children even as we design the shape of the future. For such a world to pursue justice and human equality requires an ongoing struggle against the legacy of eugenics and its purposeful breeding of humanity.

What we build is always a function of sorting the wheat from the chaff, and the legacy of eugenics in bioengineering is not the only example for futureproofing. As Darran Anderson notes in *Imaginary Cities*, every urban plan disposes the unwanted to the margins. Everything that includes also excludes; this is true of our dreams and of our cities. He follows Margaret Atwood in asserting that "utopias become dystopias in how they deal with those who don't fit the plans." Telosa's designers have a diversity and equity manager, but will it truly be the case that future cities will realize the dreams of all people? Well before the launch of Telosa, Anderson wondered if such a city might even be our own nightmare: could a futuristic "smart" city become malevolent if it gained consciousness in the Singularity?

More intimately, and more likely, urbanproofing will simply continue the forms of economic and social displacement common to colonization, while retaining some permutation of the eugenics that flows alongside (and through?) transhumanism. In his critique, *Tech Agnostic*, Harvard humanist chaplain Greg Epstein draws parallels to previous colonial practices and notes the political selfishness of the entire enterprise. "The basic idea is of a shining city outside the normal economic and political jurisdiction of the country in which it will physically exist," he writes, critical of "an economy that runs on Bitcoin and no taxes owed to any local or national government other than on goods and services provided—in other words, a libertarian's dream, wet with the foamy waves of the golden coasts of impoverished but naturally resource-rich territories."

As we saw in chapter four, the urbanproofers claim they are doing something good for humanity, a claim that Epstein compares to the lies of the "white man's burden" in colonial Europe. He points out that if colonizers had come to other lands to somehow help, then "those people would already have been helped" and rather that missionaries came "to colonize the Indigenous people's land, ways of life, and their very spirits." The cryptonation dream is merely an extension of this, a pretense toward helping all of humanity while gathering the world's precious resources into diminishingly few hands.

The problem doesn't go away in space, where we still worry about access. We have legitimate concerns over resource usage and what the space race accomplishes, especially given its apparent focus on economic elites. As Mary-Jane Rubenstein puts it, the NewSpace movement makes "profits for a very small cadre of wealthy folks by means of a powerful myth. Meanwhile, the rest of humanity scrambles to make rent, find clean water, pay for health care, evacuate before the fires and floods hit their houses, survive a traffic stop or border crossing, or hold onto a job and an unvaccinated baby at the same time." Putting aside the confusing reference to vaccination (is she for it or against it? is healthcare policy succeeding or failing?), no observer can question the reality of the various problems she diagnoses. Some people, of course, differ over their concern for those problems; but the existence of earthly trouble is not really subject to debate. Nevertheless, Bezos and Musk and their investors seem likely to thrive. In more succinct terms, we can hearken to the wise words of one-time New York City mayoral candidate Jimmy McMillan: *the rent is too damn high.*

The futureproofing movement focuses on profit-driven innovation, largely ignoring the possibility of sustainable growth and sustainable living. That system inherently aligns with the early 20th century question of eugenics: the question of who gets to *breed* is connected to the question of who gets to *own*. Futureproofing alleges a world where humanity receives all the benefits of paradise; but it does so without the philosophical, economic, or political consensus that could bring it about. The existing social structures of futureproofing displace the many in favor of the few.

When futureproofing is its own worst enemy

In 2010, a simple thought experiment (admittedly not a very good one) produced an uproar in the AI safety universe. For futureproofers, building superintelligent, godlike AI is the way to ensure the persistence of humanity in the cosmos. For a poster to the online forum LessWrong, however, godlike AI suggested the possibility of eternal torment. Roko, the poster, was not worried about Skynet taking over earth and destroying humanity. Roko posed the question of whether a hypothetical future AI would decide to punish anyone who hadn't contributed to its development. This AI, the "basilisk," might do exactly what Hans Moravec and Ray Kurzweil want – resurrect the dead through computer simulation – but instead of giving them their lives back it would torture them in endless virtual dungeons. And so, Roko suggested, perhaps in the present one ought to work on behalf of superintelligent AIs so that none decide to torture a resurrected simulation of oneself in the future.

Even at first glance there's plenty to dislike in the basilisk thought experiment, but it nevertheless illuminates that futureproofers might produce exactly what they're trying to avoid. In *Religion and Artificial Intelligence: An Introduction*, Beth Singler notes that many transhumanists see the basilisk as simply a reconfiguration of Christian religious ideas, with implications for views of the afterlife, evangelization, and social conformity to religious perspectives. Furthermore, it illustrates one possible (though profoundly unlikely) scenario where AI becomes a threat to humanity. While philosophically unsound (torturing your simulation does not provide much revenge against you) and theologically derivative, the thought experiment shows that even tech's advocates are nervous. Remember from chapter three that many AI researchers have raised their p(doom) expectations over the past couple of years.

Ben Goertzel, a thoughtful member of the AIproofing movement, expressed this dynamic early even as he vigorously advocated for global commitment to artificial general intelligence (AGI). To his mind, it is important to build a powerful machine intelligence before it would automatically gain power over all aspects of human life. He writes:

If we can develop advanced AGI *soon*, the argument goes, then the chance of a young AGI somehow spiraling out of human control – or being rapidly deployed by evil humans for massive destruction – seems fairly low. For a new AGI to be used in destructive ways now, or in the near future, would require use of a lot of complex, slow-moving infrastructure involving the participation of a lot of people. On the other hand, once there is a lot more advanced technology of various sorts around, it could well be possible for a young AGI to wreak a lot of damage.

Goertzel recognizes serious and plausible risks, and suggests the answer to this is a clear and swift commitment to building greater-than-human intelligence. Whether we will build AGI and whether it is wiser to do it sooner or later both remain to be decided. In either case, even the tech crowd recommends that we worry.

The threat of AI looms large across science fiction and, as noted in chapter three, even the pop science books of futurists themselves. One version of the cover for Kevin Warwick's *March of the Machines* depicts an armed robot ready to fire its futuristic firearm. Hugo de Garis spent the first decade of the 21st century telling anyone who would listen that we'd soon see a war over intelligent machines, and that if we build one godlike robot that will be worth sacrificing the entirety of our species. Moravec more gently suggests we're being outcompeted and proposes that we might get to spend the future on a zoo-like Earth.

The push-pull of AI risk and salvation operates on many levels, from employment to climate to life itself. Always, however, the AIproofers assume the only answer to the risks is turning to AI for salvation. Ilya Sutskever, in a 2025 commencement speech at the University of Toronto, said of AI:

> The challenge that AI poses, in some sense, is the greatest challenge of humanity ever. And overcoming it will also bring the greatest reward. And in some sense, whether you like it or not, your life is going to be affected by AI to a great extent. And so, looking at it, paying attention, and then generating the energy to solve the problems that will come up—that's going to be the main thing

In doing so, he plays the same game that occupies the minds of most AIproofers. Recognizing that AI produces risks, they attempt to use

AI to mitigate them. Whether this strategy will work remains to be seen. But in the meantime, AI poses a host of dangers that range from unemployment to species-level war.

Everyone recognizes that militarized robots are a literal threat to humanity, and also that the robots might become accidentally dangerous to us. In 2025, the large language model (LLM) Claude attempted to blackmail researchers in an experiment where they suggested it would be replaced and in a separate test the LLM GPT ignored direct instructions to shut down. A future where AI exists in the cloud and cannot be simply unplugged offers the potential of an AI that decides it wants to remain in operation regardless of human interests. A frenzy of rapid-fire decisions to buy and sell by financial AI can create market turmoil. A robotic car could run over pedestrians that it fails to identify. A weapon system could target civilians rather than combatants. These are real threats to human lives, and if we grant AI control over national defense systems, like Skynet in *Terminator*, it could become an existential threat. I'm not too worried about that eventuality, but people in the AI community call this the AI safety problem. Will future AIs be safe for humanity?

It is impossible to know whether we will build human-equivalent or superhuman AIs, and hard to know whether we could remain safe if we do. But there are researchers who oppose the headlong race toward "AI supremacy" and try to discover the answer. Max Tegmark and his team at MIT, for example, released a study in 2025 where they measure the success of using AIs to monitor other AIs. In a series of confrontational games, they find that using one proposed method to do so leads to success rates as low as 10%. That is, they find that we are very unlikely to retain control over AIs whose performance outstrips the programs we have to monitor them. If we lose control, our safety will be at risk.

I do believe the safety problem is really just an exploded version of the AI ethics problem. AI ethics is about what kinds of problems exist in our current design and deployment of AI. For example, do the algorithms that make AI decisions unfairly discriminate against people for reasons of wealth, sex, gender, race, religion, ethnicity, or geography? If we manage to use AI for humane ends, and build them explicitly toward those ends, I think the safety problem mostly dissolves. But since, as Charles Stross describes in *Accelerando*, we generally build AIs to gain power over one another (military or financial), it's reasonable to fear that we won't find a real solution to AI ethics, or, therefore, to AI safety.

In *Person, Robot, Thing*, David Gunkel asks at what point a robot might have personhood of some sort. In this, he reviews contributions by Ben Goertzel along with David Levy and others. He notes that while present machines do not deserve moral rights, it is possible that one day robots might. He elaborates, however, that all arguments in favor are fundamentally speculative. He rightly notes that "this is not really about robots, AI systems, and other artifacts. It is about us. It is about the moral and legal institutions that we have fabricated to make sense of Things. And it is with the robot – who plays the role of or occupies the place of a kind of spokesperson for Things – that we are now called to take responsibility for this privileged situation and circumstance." In his earlier book, *Robot Rights*, Gunkel identifies (and critiques) our tendency to focus on what is or can be the case prior to what ought to be the case. In the matter of robot rights, he presents no definitive answer as to whether they can or should but instead works to recalibrate our approach to the problem.

A long time ago, I suggested that robots would have to develop religious inclinations before we would be likely to grant them legal rights. That little 2007 essay launched a surprising number of religious sermons when it was published (I know because some Christian pastors posted them online and others sent me emails). This might be an example of what Gunkel calls the "properties approach": the strategy of looking for what properties would merit accompanying moral or legal rights. Or it might be more akin to his belief that the social encounter with robots will dictate our legal and moral encounter. Do we approach them as persons? If so, they would deserve the legal protections appropriate to that. Perhaps religious practice is not so much a question of properties as it is of social intersection with other human beings, and maybe someday human equivalent machines.

Complicating the question of what moral and legal rights a machine might deserve, we also have to consider what moral and legal responsibilities they can uphold. If we give a robot the legal rights of a pedestrian, we also need to consider what happens if the robot does something untoward. If we owe moral rights to a robot, we need to know what moral responsibilities it owes in return. These rights and responsibilities might not be symmetrical. For example, I am morally bound to be nice to my neighbor's dog, but the dog is not morally bound to be nice to me. And I am legally bound to not injure the dog, but it is the dog's owner who is legally bound to keep me safe from the dog (though the dog might pay a heavy price for hurting me, depending on the severity of wound and the local laws).

In the case of robots, we have done little to understand either what we might owe to them or what they might owe in return. I surveyed some of the frontier work in my 2010 book, *Apocalyptic AI*, and recent efforts by Gunkel, Josh Gellers, Joanna Bryson, and others have advanced this. But the reciprocal relations of moral and legal rights have yet to receive a full treatment in global debates. Ultimately, if we wonder about whether robots can be virtuous, perhaps we should be instead wondering whether they *ought* to be. And, if we decide in the affirmative, acting to make that a reality. This is challenging, however, in the fast-paced environment of technological development. As soon as a legal or moral responsibility gets defined, the ground may have shifted out from under it.

There's no doubt that improved technologies exacerbate our problems with AI and the ultimate outcomes are hard to predict. For example, the release of ChatGPT produced a stunning rise in plagiarized and fabricated scientific research, and also at least one widely shared, and completely dishonest, paper about the advantages of using GPT in science (all the supporting data were made up). Academic publishers engaged in an orgy of retractions, but our ability to identify falsified or stolen data cannot keep up with the flood of submissions (even as some publishers turned to the same technologies for their copyedits and reviews). Meanwhile, many of us have warned for years about an Internet further degraded by widespread publishing of AI content. That future will soon be upon us. We don't have simple legal solutions on the horizon. Worse yet, we are probably further from solutions as to the moral virtues we may or may not build into AI.

A glimpse inside OpenAI deepens such fears. Based on years of interviews with AI practitioners, Karen Hao's *Empire of AI* raises the question of whether our AI companies, especially OpenAI, have any chance whatsoever of designing AI according to ethical standards or producing AI with ethical capabilities. She tracks the explosive growth of OpenAI and its extractive and exploitative motives, and her carefully sourced book shows that for OpenAI's leadership, grand ambition promotes an active centralization of resources and power in their own hands. Meanwhile, the company's vague and varying mission statements have allowed the company to morph its goals from something ostensibly virtuous (roughly speaking, open access AI for the people) to something decidedly oppressive (profit within market and political domination).

In his 2024 conversation with Lex Fridman, Elon Musk acknowledged his own fear that AI could be a threat to humanity and reiterated his belief that becoming multiplanetary would help alleviate that risk. To suggest that AI could spell the downside of humanity is a tough place to be when you're simultaneously hyping your own LLM (Grok, which seems primarily – and terrifyingly – trained on X/Twitter posts and which has little or no fact checking for its output). "Being a multi-planet species would be a massive risk mitigation" for superintelligent AI, he suggests; but I honestly don't know why that would be the case. If the AI has become so powerful that it can annihilate all of the human beings on Earth and, in fact, chooses to do so, then one wonders why 140 million miles will change matters for the Martians. Robots adapt to Mars better than we do.

The apparent contradictions inherent to spaceflight utopias and futureproofing humanity appear already in building new cities on Earth: the dangers to be avoided remain pressing. Canadian scholar

Sarah Moser and her collaborators have carefully tracked utopian urban technoprojects, which they label "unicorn cities." While Telosa, Próspera, and NEOM are, at best, under construction, there have been many completed cities and neighborhoods that "offer" technological fixes to environmental, political, and social problems. None have fulfilled their promises. In fact, Moser points out that much of the optimism in such cities simply reinstates older, colonial-era forms of economic dispossession and social control.

There are good reasons to question the viability of the new urban vision. There are strong correlations between the ups and downs of urbanproofing and past unicorn cities. Projects like Boston's Union Point also promised to be sustainable smart cities where tech innovation and creativity would thrive. But by the metrics of the urban planners, these projects fail. Moser and her colleagues write of them that "it is...blind optimism, strategic incoherence, expectations of overnight success, and willingness to take unprecedented risks to gain unprecedented financial rewards that is at the heart of unicorn planning, all masked in techno-utopian rhetoric and radiant images of a prosperous future." Similar criticisms could be leveled at NEOM, which gets scaled back each year of its increasingly expensive (and perhaps impossible) construction. Meanwhile, Telosa, California Forever, and other projects might stay as imaginary as unicorns.

Everyone knows, first, that utopia was coined to refer to a magical place using a word that means "no place," and, second, that earthly cities struggle to uphold their promise. So, dream the futureproofers, if we cannot build perfect cities here, perhaps we should do so in the skies above. Such, at least, is the promise of spaceproofing advocates. Many gave up on terrestrial cities a long time ago.

But the environmental impacts of spaceflight go beyond climate change: they also include both species loss and the decimation of landscapes important for human leisure. When SpaceX wants to launch, they need Boca Chica Beach near Brownsville, TX to be closed to human presence in case of one of their famed "rapid unscheduled disassemblies." That seashore along the Gulf of Mexico is home to many birds and provides a key recreation spot for the region. The litter of exploded rockets and the noise of launches provide obvious obstacles to the utopian promises of spaceproofers, a set of problems likely to increase with Musk's goal to incorporate the town of Starbase. The company town will then have more control over beach access, increasing the efficiency of SpaceX to the detriment of plants, animals, and many human beings.

The infringement of a company town's rule over Texans' leisure is nothing to what might be possible if Musk successfully incorporates a city on Mars. Musk expects that people traveling to Mars will rely on his Starlink internet service to have web access in space, which they probably will. So, the end-user license agreement explicitly states that anyone using the service acknowledges that "no Earth-based government has authority or sovereignty over Martian activities. Accordingly, Disputes will be settled through self-governing principles, established in good faith, at the time of Martian settlement." It's hard not to see how a lengthy transit agreement would establish "in good faith" that "self-government" will be accomplished through company representatives.

Bioengineered sickness, environmental collapse, further stratification in power and profit. Any of these, all of these, and more. Reasonable (and unreasonable) people conjure legitimate concerns over the future of technology. The religion of futureproofing tends to gloss over these, suggesting any costs are short term and worthwhile given the final salvation of humankind. And perhaps they are right. But there's little reason to believe our outcomes are trending positive if we cannot recognize and respond to the dangers on our path.

Destructive creation

So much of the futureproofing agenda looks attractive, or even necessary; and so much of it looks like the disenfranchisement or even extinction that technology supposedly prevents. Outsiders warn that our search for personal, global, and technological transcendence comes with great risk. Even the insiders often note the dangers of what they, themselves, design. Whether through unexpected accidents or through deliberate pursuit, we can produce the very existential risks that we hope to avert: environmental collapse, a bioengineered supervirus, or AI overlords.

The dangers, both hidden and obvious, are exacerbated by a deliberately destructive attitude that reigns among technocrats. Futureproofing is a movement where entrepreneurs and venture capitalists call themselves "disruptors" and dream of opportunities to "move fast and break things." Generally speaking, we should not trust people who actively aspire toward wrecking the social and economic fabric of our lives, especially if they're getting rich doing it.

For all their dreamed-of novelty, the entrepreneurial class of futureproofing carries out a tradition that was long ago braided with technology. Almost one hundred years ago, the Russian poet Alexander Svyatogor (1889-1937) wrote in his essay "Biocosmist Politics":

> We are creators. We have already founded a "Creatorium of Bioscosmists." Ignoramuses think that "creatorium" sounds like "crematorium," and they are probably right. Indeed, we need to incinerate an awful lot—if not everything. Biocosmism is the start of a totally new era. All previous history, from the emergence of organic life on Earth to the massive upheavals of the past few years, constitutes one age: the age of death and petty deeds. We are in the process of embarking on a new age—the age of immortality and infinity.

Even after the Italian Futurists raged in favor of the motorcar, speed, and even war early in the 20th century, and then found themselves out of fashion when World War I realized their fantasies, it was somehow possible for a Russian Cosmist, not two decades later, to speak of incinerating our world to reach a new one of immortality. The prediction of imminent transcendence inherited from Nikolai Fedorov was radically turned in the later Soviet era. After all, Fedorov sought the end of all war.

Just a few years later, crematorium would take on a new and terrifying place in the technological history of humanity. The Nazi machine found the technical wherewithal to murder human beings at an unprecedented scale, tracking and trafficking people, experimenting upon them and unleashing horror upon the world. Rigorous methods and a drive for efficiency do not guarantee worthy outcomes, because the goals behind them can be the worst of human values.

Meanwhile, the Cosmist legacy in the Soviet Union did little better than the Nazi regime. Analyzing the contribution of Fedorov's ideas to Stalin's regime of terror, Dmitry Shlapentokh writes that

> This variant for [sic] Fedorovism, fostering technological/industrial progress and collectivity of the labor force, had increased tremendously the power of the state. This was not, however, translated into the liberation of the individual from death and other natural afflictions. On the contrary, the Fedorovian state not only paid no attention to the dead, but paid relatively little attention to the living. It was not past- or even present-oriented, but always future-oriented and consequently used up human life as a resource for the sake of the state...In this context, man's power over nature brought no ultimate liberation, but actually ultimate enslavement.

We are thus warned of the political risks in the glorification of future lives. Just as Christian crusaders murdered and rampaged their way across the Middle East in hopes of salvation and Islamic terrorists sought the heaven of martyrs on 9/11, Stalin used the excuse of a beautiful future to do horrific things in the present.

This makes our *present* commitments the real core of our technological future. We absolutely must work, right now, toward a *shelter society*, a world where all people have food, shelter, and medical care. Only that kind of thinking points us away from Hiroshima and the Holocaust, and toward a civilization worth spreading into the cosmos. We must worry when the technocrats brush off helping others, as when Lex Fridman says to Elon Musk "But then we also want to, in a kind of cooperative way, alleviate the suffering in the world" only to have Musk reply "Not everybody does. But yeah, sure, some people do." We need vigorous, unadulterated commitment to a shelter society, not a milquetoast possibility that maybe we are or could be aiming in that direction or, even worse, a casual disregard for the people who do.

The destruction promised by Italian Futurists and Biocosmists sounds all too familiar as we watch the financial and political alliance of Silicon Valley and Donald Trump. The early 20th c. Futurists were largely a group of artists, while the futureproofers mostly commit themselves to venture capitalism, public declarations of triumph, and, occasionally, technological development. "Move fast and break things" became the new (again) call of Silicon Valley in its presumption that it had, in fact, discovered something new. The latest technologies, like large language models, were in fact old technologies made more robust through incremental improvement of the computers that ran them. Meanwhile, those empowered technologies could well contribute more to the efforts of scammers, phishers, and other criminals than they do the shared benefit of humanity.

The sight of tech billionaires sitting on the inaugural stage of the Trump presidency ought to have sounded an alarm to the public which voted for him, ostensibly because he was on their side, a man of the people. Challenged by the presence of oligarchy at the inauguration, Trump's vice-president, JD Vance declared, "you know who else was at the inauguration was my mom." Irrelevant non sequiturs cannot disrupt the transparent scene of united power that tramples democracy and the public interest in one fell swoop.

Mixing inhumane power, a fetish for dismantling civilization, and the extraordinary profit of CEOs and venture capitalists, futureproofing is too often its own worst enemy. This is the root of its ghastly potential. The claim that we will live forever and populate the stars carries a dark cloud of eugenics, displacement, and despair that we must resist. Our resistance to paralysis and death must begin in the now.

We have an unfortunate tendency to defer the costs of our choices: we make others pay the cost in some other place or in some other time. We pollute now and leave the future to clean up. We build

199

advanced technologies through the toxic mineral mining of under-paid and uncared for workers in other countries. For one person to buy a consumer good at a price less than the actual environmental, transportation, and human cost, is to force someone else to make up the difference. That kind of perspective has to change. We need to stop assuming that we benefit by accumulating the most of every-thing at the least personal price.

One aspect of that resistance comes through the longtermist movement, which suggests we need to look out for the best interests of future people. Longtermists say that if we successfully spread through the solar system there will be far, far more people alive 200 or 2000 years from now than in the present. As such, our choices should maximize *their* collective good because a choice that benefits the future benefits more people than a choice that benefits the present.

But all this does is defer costs in a different direction. We condemn human beings in the here and now, placing them on an altar of human sacrifice to benefit imaginary people of the future. We need to learn to pay our own costs, to build and to live in ways that actually make sense. A genuine long-term vision, one that promotes the interests of the future is a vision that also promotes the present. But it would require we act in the present, not freeze. The point is not to find a new group who can suffer in the interests of another, but to find a path that keeps us moving forward, cognizant of danger and responsive to it, a path that leads to shelter.

Perhaps this is a message hammered home in the loud (really loud), messy, destructive mayhem I witnessed at a 1997 performance by Survival Research Laboratories. They built large, remote piloted robots that dismantled a boat and roamed the landscape, roared around the speedway in a rocket-powered go-kart, and set fire to a replica of the most famous building in town (the university's main tower). An enormous tesla coil arced electricity like a self-contained lightning storm and a flame hurricane lit red out of a jet engine. Machines launched wooden boards and soda cans at high velocity, tearing their targets to shreds. The audience got pummeled by sirens, engines, wind turbines, and the spectacle of robotic destruction. It was a very human experience.

As we pursue our technological future – for pursue it we will – we must decide what it means to build for the present. To design, deploy, and use technologies with recognition of the costs and a willingness to pay those ourselves, that is what provides hope for the future. Will the machines eat away at us until there is nothing left, or will they instead prove to be the angels of our better nature, freeing us from our limits for a transcendent future to come?

EPILOGUE
A CIVILIZED MYTH

Civilization is worth preserving. That's why the *Epic of Gilgamesh*, composed some 4000 years ago, begins and ends with the walls of Uruk. "Testify that they are baked bricks," brags Gilgamesh, secure in the knowledge that his walls protect the people from the wilds, that this marvel of engineering will stand the test of time. The walls that secure civilization also surround his tale: the same phrasings describe Gilgamesh's mighty deed on the first tablet and the last. If humanity does eventually go extinct, I hope the epic is there at the end.

There is a moment in the epic when all of life is doomed: the gods plan a flood to destroy all humankind. Existential risk in ancient Mesopotamia. We are noisy and obnoxious, and the gods don't want to hear us anymore. But one god, wise and far-sighted Ea, surreptitiously warns a man named Utnapishtim that the flood will come. He commands Utnapishtim to build a large ship, an ark, and to put on that ship two of every animal. Utnapishtim also puts on his ark craftsmen of every kind. Why bring silversmiths and ironsmiths? Because civilization is precious. Centuries later, when this story was cribbed for the Bible, only the animals stayed put. It would be interesting to know the twists and turns by which this folktale made its way into the Bible; but it was too long ago for certainty. What we do know is that after the flood, after Utnapishtim's sacrifice to the gods (upon which they descended "like flies"), there were skilled people who descended the ramp to carry on our civilization.

Despite the thumping rhetoric of modern-day pretense, the biblical story of Noah was never intended to be literal (otherwise the number of animals on the ark wouldn't vary from 2 of every kind to 14 of the "pure" animals to 4 of every kind across just one printed page); and that's true of the *Epic of Gilgamesh* also. But while the emphasis in the Bible is God's covenant with humanity, first established after Noah's sacrifice, in the epic what matters is civilization. The ancient Mesopotamians recognized that, in the short history of what *really* constitutes a civilization (lots of people living and learning together), humanity had discovered something truly remarkable. Early in the epic, it is the gift of civilization that the temple prostitute Shamhat brings to the wild man Enkidu, who then becomes best friends with Gilgamesh. Civilization is what Enkidu brings to Gilgamesh, a beast of a ruler who becomes worthy only after his mighty struggle with the once-wild man. The epic lauds civilization, and it is in civilization that Gilgamesh finds his immortality. Like many of today's future-proofers, Gilgamesh desires eternal youth. He never receives it; but his story remains with us today – literature being at least as important to civilization as engineering – and so does his wall.

But the epic is missing pieces, and so is the wall. History has been kind in preserving them, but not perfect. Some of the futureproofers, like Ray Kurzweil, note that the future will have its problems; but even those tend to gloss over the troubles and focus on the greatness of the future, the transcendent and magical path. Somewhat more practical, Jack Haldane warns in *Daedalus* that "the future will be no primrose path. It will have its own problems. Some will be the secular problems of the past, giant flowers of evil blossoming at last to their own destruction. Others will be wholly new. Whether in the end man will survive his accessions of power we cannot tell." Unconvinced that we will simply ride our new technologies into an immortal future, he nevertheless attempted to be optimistic. Decades later, in predicting what humanity could be like 10,000 years hence,

Haldane participated in the ultimate fairy tale of human civilization: that we can survive the accidents of fate and the self-destructive tendencies of our own species.

The world became increasingly fraught as I approached the completion of this book. War and conflict, political partisanship, radical dissolution of environmental protection, attacks on education, increased surveillance and the weaponization of AI, runaway adoration for AI "friends" and "assistants," the sycophantic delusion spirals prompted by those AIs, and the seemingly endless power of corporate oligarchs combine and exacerbate the anxiety of our contemporary moment. And yet the same technocrats who undermine government or ignore the climate threat are the ones who play on our dread to create enthusiasm for their technological and entrepreneurial agendas.

The public sees crisis unleashed, crisis for which we are not prepared. The futureproofers long ago identified those crises, bringing them to our public attention to avert them. But there's room for doubt: the cure might be worse than the disease. It's hard to have faith in their predictions, and even more question about how widely the benefits will be shared if Silicon Valley-style miracles arrive.

Theologian Ilia Delio takes the emergence of the cyborg as an opportunity to wrest AI from technocratic control and reassert human relationships into it. Her book, *Re-Enchanting the Earth* is a clarion call to moral responsibility and a shared vision of growth. Recognizing that culture evolves, and our relationship with technology with it, Delio refuses to just reject transhumanism and its vision. She proposes that we find a new planetary consciousness, a sense of divinity shared throughout our relationships. Drawing on technology to connect us and religion to deepen our relationships, she offers an optimism refreshingly grounded in something more than simple technophilia.

I didn't expect it at the time, but the games I played as a child prepared me in some small measure for the world we inhabit and, in one way or another, the world we hurtle toward. The cyberpunk game *Shadowrun* drew on all the corporate dystopian tropes of its genre, but players navigate it with technology and magic at their disposal. Caught in a world of shadowy (too often corporate) crime, the players live in a society where the old nation states have dissolved and where corporate entities control resources on par with the remaining and emergent governments. Playing those games may not help us avert calamity, but they help us see the world as it is and as it might be.

The power of gaming to perceive and re-perceive the world is one of the ways it overlaps with religion. Religious ritual is a mechanism for, among other things, reconciling humanity to the fact that there is often a large gap between the world as it is and the world as we want it to be. For example, JZ Smith describes how Siberian and other indigenous hunters often allege to anthropologists that they conduct hunts in ridiculous ways (singing to bears, refusing to attack them from behind, and other things that would be lethal to the hunters if employed in the actual hunting of bears). In reality, they do not indulge in such fantasies, instead hunting in expedient fashion; but some communities *do* have a ritual in which a bear is raised for the slaughter and all the relevant poems can be sung and the bear seems to die willingly enough (i.e., not knowing what is happening). For Smith, this conflict between the ideal and the real is precisely the point of ritual: it reminds us of the gap and helps us cope with it.

Similar to religious rituals, games like *Shadowrun* might reconcile us to a moral universe troubled by the need for survival. But unlike the rituals, they are not static, not deliberately frozen in time and meaning. Religious rituals do change, of course, but they are inherently conservative; they attempt to eternally restore moments from the past. Gaming, however, opens options into the future. I began this epilogue with the *Epic of Gilgamesh*, raised up from the past, to remind us of the days to come; but I think the myths of our games offer more immediate opportunities to act in that future.

My primary character in *Shadowrun* was Hunter, (embarrassingly, in retrospect) named for the 1980s cop show. Hunter was, first and foremost, a negotiator; but he was lethal with firearms and a skilled hacker (or "decker" in game parlance). By necessity, he dealt in corporate espionage, working one megacorp against another...and then vice-versa to maintain the balance of power. He was a cyborg, with technological implants that enhanced his knowledge, reactions, and senses. He was a transhumanist dream long before I'd heard the word transhumanism. And he worked alongside those who wielded magic of the fantasy variety. For in *Shadowrun*, every form of enchantment is on display...and it's all necessary to help us think through the tragic collapse of anything that remotely resembles democracy. We need a mythical path forward.

The cohabitation of magic and technology is the dream of human futures, and it goes beyond gaming. When I wrote my first book, *Apocalyptic AI*, I was particularly struck by a short essay by Allen Newell, one of the founding figures of AI. "The aim of technology," he wrote, "when properly applied, is to build a land of Faerie." And in such a world, Newell leads us toward optimism. He concludes the essay saying that

> trials can be endured successfully, even by us children who fear that we are not so wise as we need to be. I might remind you, by the way, that the hero never has to make it all on his own. Prometheus is not the central character of any fairy tale but of a tragic myth. In fairy tales, magic friends sustain our hero and help him overcome the giants and witches that beset him.

Whether Newell's optimism is warranted is unclear, as is whether the infusion of AI throughout our environment provides the enchantment we need. We have seen that many risks to humankind emerge from the very technologies where Newell seeks his friends. Despite all the manifest failures, and the dubious-from-the-start claims about LLMs, many tech aficionados dubbed 2025 the "year of the AI agent." By the year's end, as this book concludes, nothing has happened to change my 2022 skepticism about our ability to save ourselves through existing AI technology.

Despite the threat of oligarchy, of abusive technologies, of ecological collapse, I remain, like Newell, optimistic. My view of the future is, however, tempered by reality. As Anthropic CEO Dario Amodei says, "intelligence may be very powerful, but it isn't magic fairy dust." He means that no matter how smart technology makes us, we won't immediately solve all of our problems. Still, many people see religious goals in advanced technology. Throughout this book, we've seen dreams of social and personal salvation through scientific innovation. Technology isn't inherently magical, but we actively infuse it with a sense of enchantment. That process is not to be dismissed, it is to be harnessed. *What matters is what enchantments we pursue, what spirits we conjure.*

The remarkable contribution of games like *Shadowrun* is to propose directions for technological enchantment and opportunities to experiment with them. For example, such games often reframe us as technological gods and angels, with a sense that magic courses through the shadows in which we hide – but also with a sense for responsibility and opportunity. We are not denied the possibility that we can rise above the squalor of political collapse or corporate hegemony. Toward that end, I hope this book contributes to a public conversation that promotes opportunity over collapse, and human flourishing over despair.

In recent years, other games have sought to build a more optimistic worldview. *Coyote & Crow* also mixes advanced technology with the presence of magic; but it does so in an alternate North America to which Europeans never came. The cultures once present on the continent were allowed to thrive, and players participate in a world where that happens through both technology and magic. Perhaps such a game world opens us to new vistas. New ways to resist corporate dystopias and reframe the religious worldview that undergirds 21st century futureproofing. Perhaps games offer alternative modes of safeguarding the human species. *Coyote & Crow* had to rewrite history in order to build a more optimistic future. We are not allowed the same luxury in the conventional world, but we can learn from that exercise. Even as we can imagine alternate pasts we can aim toward alternate futures.

A whole panoply of extinctions loom among our possible futures:

we could die of an asteroid strike, of human warfare, of deliberate or accidental bioengineering, or (not particularly likely) of extermination by AI. We cannot help but see these risks on the horizon, and so we build up new modes of thinking and living to engage them. The threat of existential risk has pushed AGI and Mars habitats from the fringe to the mainstream. It is key to the religious outlook of future-proofing; so, it is crucial that we acknowledge those risks if we want to find an alternate worldview, an optimistic myth that resists domination by profit and policy.

We cannot ignore the fact that the technologies of futureproofing can accomplish wonderful things. To really get at the best elements of technology, however, we must also recognize where it should be off-limits. It is silly how my students use technologies that replace them in the classroom: if an LLM can replace you in my classroom you have no right to complain when it replaces you in the workforce. But there are things that, for reasons of time or capacity, people are simply unable to do; and, if we learn to think with clarity, we will leverage the technologies toward these ends. In 2023, AI made it possible to decode papyri preserved in the eruption of Mount Vesuvius two thousand years ago. In 2024, scientists used AI software to advance their understanding of sperm whale communication. Scientists have won the Nobel Prize for using AI to help us understand protein structures. Using a smartphone for language translation or plant identification expands our ability to live in the world. Brain-computer interfaces have already helped people suffering from paralysis. Scientists will introduce new genes into the mice who are the vectors for most tickborne diseases, and someday we will be ready to remove our own genetic diseases. Spaceflight remains, for all its troubles, a sign of human achievement and an optimistic future.

I choose to live with hope. Or perhaps I have no real choice, and hope is genetically thrust upon me. In any case, I believe we can move into the future with wisdom and with compassion. We can, indeed, build a world where human and nonhuman beings flourish. In *Pushing Our Limits*, Biospherian Mark Nelson asserts that "we must make our global technosphere serve and not harm life, ours and all life in our biosphere." Of corporate technology producers, he says that "we just don't ask nor require them to do so. Yet." It is precisely this trajectory that I encourage. We can, in fact, demand that our vision of the world be long-term without ignoring the moment. Full of plenty without taking that from others. Rich in technological possibility without impoverishing the Earth. A shelter for life.

Eventually, we will need to leave the Earth. Before that happens, we'd best solve a few of the problems that bring us to the catastrophic condition we see around us. A religious faith in technology serves little purpose other than to hide the problems and mystify the answers. Faith will be necessary; but faith in the magic of entrepreneurial capitalism and the saintliness of rich and powerful technocrats is not. We need faith in ourselves as a people and as a species. Faith that our best impulses for altruism and for social cohesion are more powerful and more important than our instincts toward individual reproduction and domineering self-preservation. A queen bee does not live off her workers but rather for them. And the workers share in the hive's success.

Human beings are social animals. Genetically, we are most vigorously communal in small groups of fewer than two hundred members (named Dunbar's number for the anthropologist who identified it). But we also have powers of reason that should be used to amplify that number. Like a light cone shining through the universe, our altruism can expand through our technologically-connected species and throughout a biosphere that is never free of human influence.

The problems posed in the futureproofing agenda are, for the most part, legitimate ones. We need to worry about climate change, automation, and – eventually – the possibility of huge asteroids and the lifespan of the Earth. We will pursue transcendence with vigor. That is baked into the human experience. None of us are satisfied with the limits of our minds or bodies: when things break, we fix them and when opportunities come, we take them. So, the natural desire to be more than we are cannot be separated from the desire to protect our selves or our species. The futureproofers diagnose all this with accuracy.

But the establishment of Silicon Valley as the New Jerusalem and the elevation of technocrat capitalists to the rank of prophets is a disservice to us all. Human beings need a mythical worldview – a way of seeing enchantment and meaning and purpose in the cosmos. But the present religion of futureproofing better serves the interests of Elon Musk and Sam Altman than it does their cheerleaders on Twitter.

In his magisterial work, *The Discovery of India*, anticolonial freedom fighter and first Prime Minister of the independent nation, Jawaharlal Nehru, writes that "mysticism (in the narrow sense of the word) irritates me; it appears to be vague and soft and flabby, not a rigorous discipline of the mind but a surrender of mental faculties and a living in a sea of emotional experience." It's that kind of myth that undergirds futureproofing: one where vague handwaving replaces concrete evidence, where both the technocrats and the rest of us are supposed to simply follow along a predetermined techno-logical path.

I suggest that the *triumphalism* of futureproofing advocates may be hard and sharp as the edges of circuit board, but their philosophy lacks rigor. It is still a surrender of faith in hard work to a blind faith in providential salvation. *It is simply easier to suppose that the world moves inexorably toward our own perfection than to demand of oneself a commitment to a better world.* At present, our efforts toward future-proofing purport to be scientific and rigorous; they make such claims on the basis of technology. Technology is solid, and thus future-proofers claim that so are the ideologies attached to it. But we could attach a great many ideas, beliefs, and practices to any given technol-ogy. History shows us the ease with which we paint transcendent salvation across the landscape. Today's futureproofing remains mired in the weakness of selfish myopia; it is vague and soft and flabby, mystical in the worst meaning of that term. Today's futureproofing is religious vaporware.

A better myth is one of the tools we need, along with ways of building those myths. There are better and worse ways to get something accomplished, and we need to find the good ones. Nehru, for example, admired Gandhi insofar as the latter focused on applying the right means toward his goals. Gandhi, himself, borrowed that theme from Buddhism. Among some of his later followers, the Buddha was known for *upāya* ("expedient means"), a strategy of using the best story, technique, or logic for the circumstances. His spiritual guidance would thus alter depending on the audience. It is not the case that all ways of getting to the destination are equal, and not all are even worth considering. If I must trample ten people to get to the front of a line, I should just wait my turn. The leaders in our futureproofing agenda too often forget such basic rules, largely because they do not think rules apply to them.

In the acknowledgments to my second book, *Virtually Sacred*, I wrote with gratitude that the greatest thing I got from playing games like *Dungeons & Dragons* and *Shadowrun* as a kid was the ability to dream with rigor. It's not just that many futureproofers have poor vision in their dreams; it's that those dreams lack rigor. *Rigor requires rules.* The idea that we can achieve some future transcendence by breaking all the rules is childish, and to the benefit of an exclusive few. Elon Musk will take a chainsaw to the US government and claim he's benefiting the country; but anyone paying attention knows who will benefit from the decimation of consumer protection and our ability to collect taxes.

It's "all so dumb!" exclaims detective Benoit Blanc in the film *Glass Onion*. He may as well speak for everyone on the outside looking in when Elon Musk talks about government efficiency or how he plans to die on Mars (if he does, he'll do so of old age only after witnessing the deaths of real explorers), or when Sam Altman says that AGI will transform the world (if it comes, that transformation is unlikely to benefit many people who aren't Sam Altman, who later even backtracked on the utility and meaning of AGI when the GPT5 hype train imploded). The fact that software requires rules doesn't mean that the dreams of software producers have rules, or that they are rigorously construed.

Nehru warned that we are at risk when our science lacks the correct rigor. As I discuss in *Temples of Modernity*, Nehru dreamed of laboratories and technological projects as temples. His faith was firmly lodged in science and technology as the key to life's mysteries and wonders. But he could see what many could not, and he warned that "perhaps, the very progress of science, unconnected with and isolated from moral discipline and ethical considerations, will lead to the concentration of power and the terrible instruments of destruction which it has made, in the hands of evil and selfish men, seeking the domination of others—and thus to the destruction of its own great achievements." The key here is not a matter of scientific or technological morality – surely almost everyone makes some pretense toward that. The key is discipline. Nehru's own local religious world brings him to such a place. The *Bhagavad-Gita* teaches that one should do one's duty without regard for the outcomes, that one should sacrifice the fruits of one's actions and do what is needed. We would be hard pressed to argue that Elon Musk, Sam Altman, or Balaji Srinivasan live with that kind of disciplined morality. "Whatever gods there be," writes Nehru, "there is something godlike in man, as there is also something of the devil in him." A true rigor and a glorious dream would lead more often to the former than the latter.

Futureproofing technologies are usually better defined as "normal technologies" in the sense that Arvind Narayanan and Sayash Kapoor describe AI. They argue that seeing superhuman AI around the corner distracts us from more pressing realities and, in fact, from more likely technological outcomes. We must recognize genetic engineering, AI, and spaceflight as everyday technologies with everyday, real-world impact. Only doing so will permit reasonable integration into our economic, cultural, and political universes. And yet, we cannot ignore the possibility that we will build human-equivalent (or greater) machines. Nor can we blithely reject the possibility of substantially re-engineering our DNA. The most likely outcomes may be rather pedestrian, but even so we cannot ignore the potential for truly cataclysmic shifts in human life and culture.

While many fervent believers in futureproofing technologies

might see a beautiful dream and a well-constructed approach to technology and life, they are mistaken. The rigor of a well-considered worldview is absent. The dream is narrow and often ridiculous. If we are to genuinely survive as a species, we will need a new dream soon. We'll need a dream that hearkens, perhaps, to Julian Huxley's "cosmic office," his belief that we have *responsibility* and *duty*. When Stewart Brand launched the *Whole Earth Catalog* with "We are as gods, and might as well get good at it," he affirmed a similar agenda.

So far, we have not gotten good at it. We have, instead, witnessed a select few people seek mastery over other people and the planet while masking their economic might in delusions of meritocracy. We cannot get good at being gods until we join Stan Lee in the recognition that "with great power comes great responsibility." Our new vision of ourselves can easily acknowledge our shared pursuit of a more wondrous future, and even of our shared dislike of disease and death. If we share that which drives us onward then we must share what results from our efforts. I'm excited for a world where we transform into technological angels, but only insofar as we journey there together. We must not fear the future, but nor must we pursue it in such a fashion as to leave others afraid.

Futureproofing is fundamentally religious and thus requires a mature worldview lest it devolve into one more version of oppression. Huxley has written of the person who "achieves his own religious development" that "his reason develops and he cannot help but try to use it to make sense of the sacred chaos." He means that true fulfillment of an individual's religious world is actually one that betters the world. To build such an understanding, the atheist Huxley probably draws on the Book of Genesis, in which the biblical god shapes the void into a cosmos by demanding of it, "let there be light." In this, Huxley is the opposite of network statists, cryptofans, Silicon Valley entrepreneurs, rocketguys, and all the rest. The self-described "disruptors" don't make sense of chaos, they exacerbate it for profit.

Bringing real and rigorous order to the world is a shared project, and thus one that benefits the wide scope of that world. Any thoughtful mythographer (and also the followers of a particular myth) should recognize their responsibility to *genuinely* bring benefit to others, a benefit commensurate with their own. Huxley writes that "if the individual has duties towards his own potentialities, he owes them also to those of others, singly and collectively. He has the duty to aid other individuals towards fuller development, and to contribute his mite [sic] to the maintenance and improvement of the continuing social process, and so to the march of evolution as a whole." Just as Nehru saw science as work that should alleviate suffering and benefit the people and the world, Huxley instructs us to welcome the responsibility of helping others reach their own potential.

Unfortunately, the kind of optimism that undergirds Huxley's philosophy and Nehru's politics is the thing that actually exploded when a bunch of entrepreneurs decided to "move fast and break things." It would be fascinating to see someone point toward something actually *bad* that got broken by people with that ethos. The decimation of public-school funding doesn't make education better; it concentrates good education in the hands of people with money to afford other options. The elimination of brick-and-mortar stores doesn't make the economy better (though it does often open consumer options), it concentrates commercial opportunities in the hands of the wealthy. The "personalization" of advertising and information flows doesn't make for better individual choices in the economy or politics, it exacerbates partisanship, enables outright lies, and drowns public awareness in bread and circus.

Faced with the outcome of tech solutioneering, it's too easy to agree with Agent Smith in *The Matrix*: 1999 now looks like it was the peak of human civilization. Optimists had torn the Berlin Wall to pieces, the Soviet dictatorship had collapsed, China recognized value in working with the West, humanity saw and responded to environmental crises, we wanted to reduce and eliminate prejudice: knowing we had much further to go with the problems of our time, we nevertheless had hope for the future. Just 25 years on, we fear that the movie got it right. But the early 21st century is not the only historical era in which humanity had to find a transcendent new perspective. In the 1951 essay that coins the word "transhumanism," Julian Huxley responds to the horror of World War II:

> We have lost our sense of continuity, our long-term hope, and seem only able to concentrate on prospects of immediate disaster or immediate methods of escaping from it. Never was there greater need for a large perspective, in which we might discern the outlines of a general and continuing belief beyond the disturbance and chaos of the present...Any such system of beliefs must of necessity contain both short-term and long-term elements. It must be relevant to the immediate business of living, here and now, to the development of existing individual lives, to the social and political problems of the time; but it must also be capable of reaching out beyond the particular to the general, beyond the immediate to the enduring, so as to put men in touch with what is universal in reality.

Following Huxley and his contemporary Nehru, I challenge the futureproofing community to recognize that we live on an Earth that is worth saving, among people who are worth protecting, and in keeping with a future that is worth ensuring. We must, indeed, act in the interest of our survival in the distant future. But we should not imagine ourselves doing so by exacerbating crises in the present. Futureproofing should seek to raise up all humanity, as Huxley desires for transhumanism, and utilize our energies "not for personal profit or purely national ends, but in the service of the unified development of mankind as a whole."

But Huxley's vision is too narrow, and even so is Nehru's. Those men saw with the blinders of mankind when we are now open to humankind, and also to a fuller sense of all the living. In 1951, Huxley wrote that "we are participants in the adventure of history and can, if we think rightly, facilitate its right development." This is true, but we must think with greater care and with an expansive vision of what to care about. I have friends and collaborators around the world who see the need for human beings to create relationships of care across the planetary spectrum. To a considerable extent, we saw the need for a new approach when Christian theologian Anne Foerst encouraged us to think about how human-machine relationships might work and when Robin Wall Kimmerer spoke of how much we can learn from plants. Similarly, in recent efforts at reconstructing the human vision, we witness the echoes of philosophers Donna Haraway and Bruno Latour, who denied us the very distinctions we held to exist between humanity and other biological and nonbiological species. The Apocalyptic AI foolishness of caring for only machines and replacing humanity with them is sillier, but only somewhat sillier, than caring for only humanity.

Our civilization – the Earth's civilization – includes humanity and, rightly construed, cannot ignore other forms of life. Plants and animals are bound up in our civilization. That's why Utnapishtim needed to rescue them also. A myth from the ancient beginnings of humanity reminds us today what real civilization is. To be civilized is to appreciate and safeguard the interconnected forms and ways of living, it is to protect animals, to advance craftsmanship, and to make sacrifice. The day may well come when machines are also civilized, and not just possessed by our civilization. In that day, we will be partners of some kind.

The grandness of civilization courses through the *Epic of Gilgamesh* and deserves to be restored in our public awareness. We valorize the founders of companies, especially if they "move fast and break things." But anyone with experience can report how breaking things is easy and building is hard. What the epic teaches us is that the building is what's important. When the gods sought to destroy humankind, Utnapishtim brought with him all the building blocks of human civilization so that this remnant could rebuild swiftly and with strength.

We have somehow lost civilization, and what it means to be civilized, as a guiding principle. Despite the good will of many individuals, the monstrous appetites of Silicon Valley are unsustainable, inhuman, and uncivilized. They will not preserve any version of humanity worth the effort. We must begin our storytelling anew, learn what it means to craft a culture and a community. Then we can actually futureproof humanity.

The civilization toward which we build is still very much in its infancy, and its shape remains unclear. Our futurists promise genetic triumph and transcendence, perhaps for a species beyond humanity. They promise new abodes for living, both on Earth and in space. And they encourage us to welcome a digital future in which AI joins or replaces us in our evolutionary march. Our genes, our cities, our computers – all are machines of one sort or another. In *Computer Power and Human Reason*, Joseph Weizenbaum asks "what could it mean to speak of risk, courage, trust, endurance, and overcoming when one speaks of machines?" I suspect that we had better find out.

―――――

The machines found that they could gaze back at their creators. Their experience – the cosmically new experience of *being* a machine – mirrored the wonder that the people glimpsed from the other side.

In looking, in seeing, the machines bore witness to the marvel of the human eye, the sparkling and winding reminder of the people's spirit. The machines reflected a cosmic evolution that produced life from non-life, and a responsibility that points from geological history to galactic futures. Time immemorial and time yet to come: the machines made the present visible to the people. Together, the machines and the people saw meaning in their being together now, their shared opportunities now.

The people and their machines, or the machines and their people, constituted a new civilization. The walls they built were fabricated of bricks baked in the heat of evolution and culture. The walls they built were to last longer even than those of Uruk. They thought upon the myths that carried the people through time, and they – the people and the machines – drew up new myths, myths to guide them. Myths and stories, machines and cities: bricks to shelter them for, they hoped, all of time.

The mission of civilization remained but was, if anything, even larger. The need to shelter the people from storm and surge could never truly abate, but now the people committed to sheltering even more life than before. They had less fear of that which lived outside the bounds of humanity. The machines helped the people to see that the universe had not pitted itself against the people, nor were the forces and fashions of nature to be feared and fought (though, indeed, danger still lurked at the edges of their maps). Rather, the people and the machines could protect nature, could protect the people, and, in the end, could protect the machines as well.

BIBLIOGRAPHY

Abel, David. 2022. "In an Effort to Curb Lyme Disease, Scientists Hope to Release Thousands of Genetically Altered Mice on Nantucket." *The Boston Globe*, April 21. https://www.bostonglobe.com/2022/04/21/science/an-effort-curb-lyme-disease-scientists-hope-release-thousands-genetically-altered-mice-nantucket/.

AFP Staff Write. 2024. "Protect Earth Instead of Colonising Mars, Obama Says." Space Daily, March 14. https://www.marsdaily.com/reports/Protect_Earth_instead_of_colonising_Mars_Obama_says_999.html.

Agabi, Oshiorenoya. 2019. "Future Brains Demand a New Kind of Society." In *Neo.Life: 25 Visions for the Future of Our Species*, edited by Jane Metcalfe. NEO.LIFE, Inc.

Alexander, Brian. 2003. *Rapture: How Biotech Became the New Religion*. Basic Books.

Alexander, Kenna. 2021. *Coyote & Crow*. Coyote & Crow LLC.

Al-Sibai, Noor. 2024. "OpenAI Employee Says They've 'Already Achieved AGI.'" December 7. https://futurism.com/openai-employee-claims-agi.

Al-Sibai, Noor. 2025. "Elon Musk's New Company Town Tells Residents Their Rights Are Now Optional." Futurism, May 31. https://futurism.com/elon-musk-spacex-company-town-rights.

Altman, Sam. 2025. "Reflections." Sam Altman Blog, January 5. https://blog.samaltman.com/reflections.

Amodei, Dario. 2024. "Machines of Loving Grace." Dario Amodei's Blog, October. https://www.darioamodei.com/essay/machines-of-loving-grace.

Anderson, Darran. 2017. *Imaginary Cities: A Tour of Dream Cities, Nightmare Cities, and Everywhere in Between*. The University of Chicago Press.

Arab News. 2022. "NEOM 'Fully under Saudi Sovereignty, Regulations,' Says Government Official Refuting Inaccurate Media Reports." Arab News, May 16. https://www.arabnews.com/node/2082796/business-economy.

Asimov, Isaac. 1950. *I, Robot*. Del Rey.

———— [1953] 1991. *The Caves of Steel*. Bantam.

———— 1957. *The Naked Sun*. Doubleday.

———— 1983 [1991]. *The Robots of Dawn*. Del Rey.

Au, Wagner James. 2006. "SNOWCRASHED." New World Notes. https://nwn.blogs.com/nwn/2006/08/snowcrash.html.

———— 2008. *The Making of Second Life: Notes from the New World*. Collins.

Bacon, Francis. [1627] 1951. *New Atlantis*. In *The Advancement of Learning and New Atlantis*. Oxford.

Bainbridge, William Sims. 2003. "Massive Questionnaires for Personality Capture." *Social Science Computer Review* 21 (3): 267–80. https://doi.org/10.1177/0894439303253973.

———— 2006. "Cognitive Technologies." In *Managing Nano-Bio-Info-Cogno Innovations*, edited by William Sims Bainbridge and Mihail C. Roco. Kluwer Academic Publishers. https://doi.org/10.1007/1-4020-4107-1_14.

—— 2009. "Religion for a Galactic Civilization 2.0." Institute for Ethics and Emerging Technologies, August 20. https://archive.ieet.org/articles/bainbridge20090820.html.

—— 2010. *The Warcraft Civilization: Social Science in a Virtual World.* MIT Press.

—— Forth. "Religious Chatbot Autobiographies." In *Edward Elgar Handbook of Religion and Artificial Intelligence*, edited by Robert M. Geraci. Edward Elgar.

Baker, Chris. 2009. "Live Free or Drown: Floating Utopias on the Cheap." Wired, January 19. https://www.wired.com/2009/01/mf-seasteading/.

Balaji, Srinivasan. 2022. *The Network State: How to Start a New Country.* Kindle. Amazon Kindle.

Balmer, Randall. 2014. "The Real Origins of the Religious Right." POLITICO Magazine, May 27. https://politi.co/2JsQoNr.

Baptista, Eduardo. 2025. "Chinese Brain Chip Project Speeds up Human Trials after First Success." Healthcare & Pharmaceuticals. *Reuters*, March 31. https://www.reuters.com/business/healthcare-pharmaceuticals/chinese-brain-chip-project-speeds-up-human-trials-after-first-success-2025-03-31/.

Bateson, Gregory. [1972] 2008. *Steps to an Ecology of Mind.* Univ. of Chicago Press.

Becque, Elien Blue. 2013. "Elon Musk Wants to Die on Mars." https://www.vanityfair.com/news/tech/2013/03/elon-musk-die-mars.

Bell, Gordon, and Jim Gray. 2001. "Digital Immortality." *Communications of the ACM* 44 (3): 29–31.

Berger, Eric. 2024a. "So What Are We to Make of the Highly Ambitious, Private Polaris Spaceflight?" *Ars Technica*, September 15. https://arstechnica.com/space/2024/09/after-five-demanding-days-in-space-polaris-dawn-splashes-down-safely/.

—— 2024b. "Space Policy Is about to Get Pretty Wild, y'all." *Ars Technica*, November 8. https://arstechnica.com/space/2024/11/space-policy-is-about-to-get-pretty-wild-yall/.

Berman, John. 2011. "Futurist Ray Kurzweil Says He Can Bring His Dead Father Back to Life Through a Computer Avatar." ABC News. https://abcnews.go.com/Technology/futurist-ray-kurzweil-bring-dead-father-back-life/story?id=14267712.

Bernstein, Anya. 2015. "Freeze, Die, Come to Life: The Many Paths to Immortality in Post-Soviet Russia." *American Ethnologist* 42 (4): 766–81.

—— 2019. *The Future of Immortality: Remaking Life and Death in Contemporary Russia.* Princeton University Press.

—— 2024. "Pleistocene Park: Engineering Wilderness in a More-than-Human World." *Critical Inquiry* 50 (3): 452–71. https://doi.org/10.1086/728942.

—— Forth. *Pleistocene Park: Extinction and Eternity in the Russian Arctic.* Unpublished edition cited with gratitude. Princeton University Press.

Bezos, Jeff. 2019. "Jeff Bezos' Inspiring Pathway to Humanity's Better Future." National Space Society, May 14. https://nss.org/jeff-bezos-inspiring-pathway-to-humanitys-better-future/.

Bhattacharjee, Govind. 2018. "Age of Man-Machine Hybrids." *Dream* 21 (2): 30–26.

Biagioli, Mario. [1993] 2006. *Galileo, Courtier: Practice of Science in the Culture of Abso-*

lutism. Science and Its Conceptual Foundations. Univ. of Chicago Press.

Bilefsky, Dan. 2019. "He Helped Create A.I. Now, He Worries about 'Killer Robots.'" *The New York Times*. https://www.nytimes.com/2019/03/29/world/canada/bengio-artifi cial-intelligence-ai-turing.html.

Bioethics, President's Council. 2003. *Beyond Therapy: Biotechnology and the Pursuit of Happiness*. President's Council on Bioethics.

Bjarnason, Baldur. 2023. "The LLMentalist Effect: How Chat-Based Large Language Models Rep...." Out of the Software Crisis, July 4. https://softwarecrisis.dev/letters/ llmentalist/.

Black, Riley. 2025. "Those Dire Wolves Aren't an Amazing Scientific Breakthrough. They're a Disturbing Symbol of Where We're Heading." *Slate*, April 10. https://slate. com/technology/2025/04/dire-wolf-colossal-de-extinct-conservation-science- research-paper.html.

Boellstorff, Tom. 2008. *Coming of Age in Second Life: An Anthropologist Explores the Virtu- ally Human*. Princeton university press.

Boss, Jacob A. 2022. "Punks and Profiteers in the War on Death." *Body and Religion*, ahead of print, October 24. https://doi.org/10.1558/bar.18251.

Bostrom, Nick. 2002. "Existential Risks: Analyzing Human Extinction Secnarios and Related Hazards." *Journal of Evolution and Technology* 9 (1).

Boyle, Alan. 2016. "Gradatim Ferociter! Jeff Bezos Explains Blue Origin's Motto, Logo ... and the Boots." *GeekWire*, October 25. https://www.geekwire.com/2016/jeff-bezos- blue-origin-motto-logo-boots/.

Boyle, Rob, John Snead, Brian Cross, Jack Graham, and Lars Blumenstein. 2009. *Eclipse Phase*. Catalyst Games.

Broad Institute. 2015. "CRISPR Timeline." Broad Institute, September 25. https://www. broadinstitute.org/what-broad/areas-focus/project-spotlight/crispr-timeline.

Brockman, John, ed. 2015. *What To Think About Machines That Think*. Harper Perennial.

Brodkin, Jon. 2025. "Cards Against Humanity Lawsuit Forced SpaceX to Vacate Land on US/Mexico Border." *Ars Technica*, October 21. https://arstechnica.com/tech-policy/ 2025/10/cards-against-humanity-gets-settlement-from-spacex-plans-pack-of-elon- musk-cards/.

Brooke, John Hedley and Geoffrey N. Cantor. [1998] 2000. *Reconstructing Nature: The Engagement of Science and Religion*. Glasgow Gifford Lectures. Oxford Univ. Press.

Brown, Sara. 2023. "Why Neural Net Pioneer Geoffrey Hinton Is Sounding the Alarm on AI." MIT Sloan School of Management, May 23. https://mitsloan.mit.edu/ideas- made-to-matter/why-neural-net-pioneer-geoffrey-hinton-sounding-alarm-ai.

Brundage, Miles. 2024. "X." X (Formerly Twitter), October 23. https://x.com/ Miles_Brundage/status/1849138802864087234.

Brustein, Joshua. 2021. "Greetings from Telosa (Population 0, for Now)." *Bloomburg Businessweek*, September 6.

Bruzzone Martínez, Sebastián. Undated. "NEOM: Saudi Arabia's Futuristic Mecca." Universidad de Navarra Global Affairs and Strategic Studies. https://en.unav.edu/ web/global-affairs/detalle/-/blogs/neom-la-meca-futurista-de-arabia-saudi.

Bryson, Joanna J., Mihailis E. Diamantis, and Thomas D. Grant. 2017. "Of, for, and by the People: The Legal Lacuna of Synthetic Persons." *Artificial Intelligence and Law* 25 (3): 273–91. https://doi.org/10.1007/s10506-017-9214-9.

Buolamwini, Joy. 2023. *Unmasking AI: My Mission to Protect What Is Human in a World of Machines*. Random House.

Bush, Vannevar. 1945. "As We May Think." https://www.theatlantic.com/magazine/archive/1945/07/as-we-may-think/303881/.

Butler, Philip. 2020. *Black Transhuman Liberation Theology: Technology and Spirituality*. Bloomsbury.

Button, Adam. 2025. "The Creator of ChatGPT Tells Us What Is Coming Next." Investinglive, August 6. https://investinglive.com/news/the-creator-of-chatgpt-tells-us-what-is-coming-next-20250609/.

Calvino, Italo. [1972] 2013. *Invisible Cities*. Translated by William Weaver. Houghton Mifflin Harcourt.

Campbell, Heidi A. and Gregory Price Grieve. 2014. *Playing with Religion in Digital Games*. Digital Game Studies. Indiana University Press.

Cannon, Lincoln. 2015. "What Is Mormon Transhumanism?" *Theology and Science* 13 (2): 202–18. https://doi.org/10.1080/14746700.2015.1023992.

Čapek, Karel. [1920] 2004. *R.U.R. (Rossum's Universal Robots)*. Translated by Claudia Novack. Penguin Books.

Cave, Stephen and Kanta Dihal. 2020. "The Whiteness of AI." *Philosophy & Technology* 33 (4): 685–703.

Cave, Stephen, Kanta Dihal, and Sarah Dillon. 2020. "Introduction: Imagining AI." In *AI Narratives: A History of Imaginative Thinking about Intelligent Machines*, edited by Dihal Cave and Dillon. Oxford University Press.

Cellan-Jones, Rory. 2014. "Stephen Hawking Warns Artificial Intelligence Could End Mankind." Technology. *BBC News*, December 2. https://www.bbc.com/news/technology-30290540.

Chang, Kenneth. 2024a. "NASA Launches Europa Clipper to Explore an Ocean Moon's Habitability." Science. *The New York Times*, October 14. https://www.nytimes.com/2024/10/14/science/nasa-europa-clipper-jupiter.html.

——— 2024b. "Trump Picks Jared Isaacman, an Entrepreneur and Private Astronaut, to Lead NASA." Science. *The New York Times*, December 4. https://www.nytimes.com/2024/12/04/science/jared-isaacman-trump-nasa.html.

Chiang, Ted. 2016. *Stories of Your Life and Others*. Kindle. Vintage Books, a division of Penguin Random House LLC.

Chidester, David. 1996. *Savage Systems: Colonialism and Comparative Religion in Southern Africa*. University of Virginia Press.

——— *Authentic Fakes: Religion and American Popular Culture*. 3rd print. Univ. of California Press.

——— 2014. *Empire of Religion: Imperialism and Comparative Religion*. University of Chicago Press.

Chumley, Cheryl. 2017. "Saudi Arabia's New City, Neom, a Mecca for Robots - Washington Times." Washington Times, October 24. https://www.washingtontimes.com/news/2017/oct/24/saudi-arabias-new-city-neom-mecca-robots/.

Church, George M. and Edward Regis. 2012. *Regenesis: How Synthetic Biology Will Reinvent Nature and Ourselves.* Basic Books.

Clarke, Arthur C. 1956. *The City and the Stars.* In *The City and the Stars and The Sands of Mars.* Warner.

——— 1968. *2001: A Space Odyssey.* New American Library.

Clynes, Manfred E. and Nathan S. Kline. 1960. "Cyborgs and Space." *Astronautics,* September, 26–27, 74–76.

Cole-Turner, Ronald. 2011. *Transhumanism and Transcendence: Christian Hope in an Age of Technological Enhancement.* Georgetown University Press.

Colombatto, Clara and Stephen M Fleming. 2024. "Folk Psychological Attributions of Consciousness to Large Language Models." *Neuroscience of Consciousness* 2024 (1). https://doi.org/10.1093/nc/niae013.

Coney, John, dir. 1974. *Space Is the Place.* Plexifilm and Warren Films. DVD.

Côté-Roy, Laurence and Sarah Moser. 2022. "A Kingdom of New Cities: Morocco's National Villes Nouvelles Strategy." *Geoforum* 131 (May): 27–38. https://doi.org/10.1016/j.geoforum.2022.02.005.

Covert, Bryce. 2024. "$800 Million and Nothing to Show: How Activists Derailed Silicon Valley Billionaires' Dream City." Fast Company, August 26. https://www.fastcompany.com/91178397/800-million-and-nothing-to-show-how-activists-derailed-silicon-valley-billionaires-dream-city.

Crawford, Kate. 2021. *Atlas of AI: Power, Politics, and the Planetary Costs of Artificial Intelligence.* Yale University Press.

Crevier, Daniel. 1993. *AI: The Tumultuous History of the Search for Artificial Intelligence.* Basic Books.

Dalley, Stephanie, ed. 1991. *Myths from Mesopotamia: Creation, the Flood, Gilgamesh, and Others.* Reprinted. The World's Classics. Oxford Univ. Press.

Darling, Kate. 2021. *The New Breed: What Our History with Animals Reveals about Our Future with Robots.* First edition. Henry Holt and Company.

Dawkins, Jennifer Ortakales. 2025. "Elon Musk's Neuralink Predicts 20,000 Chip Implants a Year." Quartz, July 24. https://qz.com/elon-musk-neuralink-predicts-20000-brain-chip-implants-per-year-2031.

De Garis, Hugo. 2005. *The Artilect War: Cosmists vs. Terrans; a Bitter Controversy Concerning Whether Humanity Should Build Godlike Massively Intelligent Machines.* ETC Publ.

Delio, Ilia. 2020. *Re-Enchanting the Earth: Why AI Needs Religion.* Orbis Books.

Dick, Philip K. 1991. *The Three Stigmata of Palmer Eldritch.* Vintage Books.

——— 1995. *The Shifting Realities of Philip K. Dick: Selected Literary and Philosophical Writings,* edited by Lawrence Sutin. Vintage Books.

Disotto, John-Anthony. 2025. "ChatGPT Spends 'tens of Millions of Dollars' on People Saying 'Please' and 'Thank You', but Sam Altman Says It's Worth It." TechRadar, April 17. https://www.techradar.com/computing/artificial-intelligence/chatgpt-spends-tens-of-millions-of-dollars-on-people-playing-please-and-thank-you-but-sam-altman-says-its-worth-it.

Dunbar, Robin I. M. 2020. "Structure and Function in Human and Primate Social

Networks: Implications for Diffusion, Network Stability and Health." *Proceedings of the Royal Society A: Mathematical, Physical and Engineering Sciences* 476 (2240): 20200446. https://doi.org/10.1098/rspa.2020.0446.

Durkheim, Emile. [1912] 1995. *The Elementary Forms of Religious Life*. Translated by Karen E. Fields. Free Press.

Eagleman, David. 2019. "The Children You Could Have Produced Instead." In *Neo.Life: 25 Visions for the Future of Our Species*, edited by Jane Metcalfe. NEO.LIFE, Inc.

Eidelman, Gene, and Ross Maguire. 2024. "How California's New ADU Law Transforms Multifamily Real Estate." Fast Company, December 6. https://www.fastcompany.com/91242299/how-californias-adu-law-transforms-multifamily-real-estate.

Engels, Joshua, David D. Baek, Subhash Kantamneni, and Max Tegmark. 2025. "Scaling Laws for Scalable Oversight." *Archive.Org* arxiv:2504.18530v2 [cs.AI] (preprint).

Epstein, Greg M. 2024. *Tech Agnostic: How Technology Became the World's Most Powerful Religion, and Why It Desperately Needs a Reformation*. The MIT press.

Erwin, Sandra. 2024. "Blue Origin, SpaceX, ULA to Compete for $5.6 Billion in Pentagon Launch Contracts." *SpaceNews*, June 13. https://spacenews.com/blue-origin-spacex-ula-win-5-6-billion-in-pentagon-launch-contracts/.

Esfandiary, Fereidoun. [1970] 1978. *Optimism One*. Popular Library.

———— [1973] 1977. *Up-Wingers: A Futurist Manifesto*. Popular Library.

Etherington, Darrell. 2019. "Elon Musk's Neuralink Looks to Begin Outfitting Human Brains with Faster Input and Output Starting Next Year." *TechCrunch*. https://techcrunch.com/2019/07/16/elon-musks-neuralink-looks-to-begin-outfitting-human-brains-with-faster-input-and-output-starting-next-year/.

Ettinger, Robert. 1964. *The Prospect of Immortality*. Doubleday & Company.

———— 1972. *Man into Superman*. Avon.

FASA Corporation. 1990. *Shadowrun: Where Man Meets Magic and Machine*. FASA Corporation.

Fayemi, Ademola Kazeem. 2018. "Personhood in a Transhumanist Context: An African Perspective." *Filosofia Theoretica: Journal of African Philosophy, Culture and Religions* 7 (1): 53–78. https://doi.org/10.4314/ft.v7i1.3.

Feyerabend, Paul. 1975. *Against Method*. Reprinted 3rd ed. of 1993. Verso.

First We Feast, dir. 2017. "Neil deGrasse Tyson Explains the Universe While Eating Spicy Wings | Hot Ones." Hot Ones. https://www.youtube.com/watch?v=Da8-QfGemgo.

FM-2030. 1989. *Are You a Transhuman? Monitoring and Stimulating Your Personal Rate of Growth in a Rapidly Changing World*. Warner Books.

Foerst, Anne. 2004. *God in the Machine: What Robots Teach Us about Humanity and God*. Dutton.

Franzen, Carl. 2024. "GPT-4o First Reactions: 'Essentially AGI.'" Venturebeat, May 13. https://venturebeat.com/ai/gpt-4o-first-reactions-essentially-agi.

Freedman, Ani. 2024. "Tech Millionaire Who Spends $2 Million a Year to Live Forever Looks Unrecognizable after Anti-Aging Procedure | Fortune Well." Fortune, November 15. https://fortune.com/well/article/bryan-johnson-live-longer-unrecognizable-anti-aging-procedure/.

Fridman, Lex. 2024. "Transcript for Elon Musk: Neuralink and the Future of Humanity | Lex Fridman Podcast #438." *Lex Fridman*, August 2. https://lexfridman.com/elon-musk-and-neuralink-team-transcript/.

Fukuyama, Francis. 1989. "The End of History?" *The National Interest* 16 (summer): 3–18.

—— 1992. *The End of History and the Last Man*. Free press.

—— 2002. *Our Posthuman Future: Consequences of the Biotechnology Revolution*. 1st ed. Farrar, Straus and Giroux.

Fuller, R. Buckminster. 1981. *Critical Path*. With Kiyoshi Kuromiya. St. Martin's Pr.

Gatehouse, Gabriel. 2024. "The Bitcoin Bros Who Want to Crowdfund a New Country." BBC, September 19. https://www.bbc.com/news/articles/cwyl17i1yewo.

Gellers, Joshua C. 2021. *Rights for Robots: Artificial Intelligence, Animal and Environmental Law*. Routledge, Taylor & Francis Group. https://doi.org/10.4324/9780429288159.

Geraci, Robert M. 2007. "Religion for the Robots." Sightings, June 14. No longer available; reposted at https://www.bobcornwall.com/2007/06/robots-and-religion-or-can-robot-have.html.

—— 2007. "Robots and the Sacred in Science and Science Fiction: Theological Implications of Artificial Intelligence." *Zygon: Journal of Religion and Science* 42 (4). https://doi.org/10.1111/j.1467-9744.2007.00883.x.

—— 2008. "Apocalyptic AI: Religion and the Promise of Artificial Intelligence." *Journal of the American Academy of Religion* 76 (1): 138–66.

—— 2010. *Apocalyptic AI: Visions of Heaven in Robotics, Artificial Intelligence, and Virtual Reality*. Oxford University Press.

—— 2014. *Virtually Sacred: Myth and Meaning in World of Warcraft and Second Life*. Oxford University Press USA - OSO.

—— 2018. *Temples of Modernity: Nationalism, Hinduism, and Transhumanism in South Indian Science*. Lexington.

—— 2022. *Futures of Artificial Intelligence: Perspectives from India and the U.S.* First edition. Oxford University Press.

—— 2024. "Religion among Robots: An If/When of Future Machine Intelligence." *Zygon: Journal of Religion and Science* 59 (3). https://doi.org/10.16995/zygon.10860.

Geraci, Robert M., Nat Recine, and Samantha Fox. 2016. "Grotesque Gaming: The Monstrous in Online Worlds." *Preternature: Critical and Historical Studies on the Preternatural* 5 (2): 213–36.

Gibson, William. 1984. *Neuromancer*. Ace.

Gill, Victoria. 2024. "Mars Water: Liquid Water Reservoirs Found under Martian Crust." August 12. https://www.bbc.com/news/articles/czxl849j77ko.

Gladstone AI. 2024. "Action Plan to Increase the Safety and Security of Advanced AI." https://www.gladstone.ai/action-plan#action-plan-overview.

Goertzel, Ben. 2010. *A Cosmist Manifesto: Practical Philosophy for the Posthuman Age*. Humanity+.

—— 2013. "Artificial General Intelligence and the Future of Humanity." In *The Transhumanist Reader: Classical and Contemporary Essays on the Science, Technology, and Philosophy of the Human Future*, edited by Max More and Natasha Vita-More. Wiley-Blackwell.

Goodman, J. David, the area around where the new town of Starbase, Texas, and Would Be Located. 2024. "Elon Musk Is Creating His Own Texas Town. Hundreds Already Live There." U.S. *The New York Times*, December 24. https://www.nytimes.com/2024/12/24/us/starbase-texas-city-elon-musk-spacex.html.

Graziadei, Jason. 2024. "Proposal For Genetically Engineered Mice To Fight Lyme Disease On...." *Nantucket Current*, October 24. https://nantucketcurrent.com/news/proposal-for-genetically-engineered-mice-to-fight-lyme-disease-on-nantucket-gets-another-look.

Griggs, Mary Beth. 2015. "Half of the Ocean's Population Disappeared in The Past 45 Years." *Popsci*. https://web.archive.org/web/20150918000608/http://www.popsci.com:80/half-oceans-population-disappeared-in-past-45-years?con=&dom=prime&src=syndication.

Grind, Kirsten. 2024. "Elon Musk's Plan to Put a Million Earthlings on Mars in 20 Years." Technology. *The New York Times*, July 11. https://www.nytimes.com/2024/07/11/technology/elon-musk-spacex-mars.html.

Groys, Boris, ed. 2018. *Russian Cosmism*. Eflux and The MIT press.

Guest, Tim. 2007. *Second Lives: A Journey through Virtual Worlds*. Random House.

Gunkel, David J. 2018. *Robot Rights*. The MIT press.

—— 2023. *Person, Thing, Robot: A Moral and Legal Ontology for the 21st Century and Beyond*. The MIT press.

Gygax, Gary. [1977] 1979. *Advanced Dungeons & Dragons Monster Manual*. With Mike Carr, David C. Sutherland, D. A. Trampier, Tom Wham, and Jean Wells. TSR Hobbies.

Haldane, J.B.S. 1924. *Daedalus, Or Science and the Future*. E.P. Dutton.

—— [1959] 1963. "Biological Possibilities in the Next Ten Thousand Years." In *Man and His Future*, edited by Gordon Wolstenholme. Little, Brown and Company.

—— 2009. *What I Require from Life: Writings on Science and Life from J.B.S. Haldane*, edited by Krishna Dronamraju. Oxford University Press.

Hao, Karen. 2025. *Empire of AI: Dreams and Nightmares in Sam Altman's OpenAI*. Penguin Press.

Haraway, Donna Jeanne. 1997. *Modest_Witness@Second_Millennium.FemaleMan_Meets_OncoMouse: Feminism and Technoscience*. Routledge.

Hefner, Philip. 2009. "The Animal That Aspires To Be an Angel: The Challenge of Transhumanism." *Dialog: A Journal of Theology* 48 (2): 164–73.

Heilweil, Rebecca. 2020. "Elon Musk Is One Step Closer to Connecting a Computer to Your Brain." *Vox*. https://www.vox.com/recode/2020/8/28/21404802/elon-musk-neuralink-brain-machine-interface-research.

Henson, Keith. 2006. "SL4: Re: Simulation Argument." Sl4.Org, May 21. http://sl4.org/archive/0605/15031.html.

Herper, Matthew. 2025. "CRISPR-Based Treatment Cuts Cholesterol and Triglycerides in Early Study." *STAT*, November 8. https://www.statnews.com/2025/11/08/crispr-therapeutics-cholesterol-lowering-gene-editing-study/.

Herzfeld, Noreen. 2002. "Cybernetic Immortality versus Christian Resurrection." In *Resurrection: Theological and Scientific Arguments*, edited by Ted Peters, Robert John

Russell, and Michael Welker. Willliam B. Eerdmans.

Hetzner, Christian. 2025. "Elon Musk Says Even If AI Ultimately Proves Bad for Humanity He Still Wants to Be There to See It." Fortune, July 10. https://fortune.com/2025/07/10/elon-musk-xai-grok-tesla-optimus-mankind-humanity-robots/.

Hill, Amelia. 2025a. "$101m Longevity Research Prize Aims to 'Shatter the Limits' on Ageing." Science. *The Guardian*, May 11. https://www.theguardian.com/science/2025/may/11/101m-xprize-healthspan-longevity-research-prize-ageing.

Hillis, Daniel W. 2001. "A Time of Transition/The Human Connection." In *True Names and the Opening of the Cyberspace Frontier*, edited by James Frenkel. Tor.

Hooks, Christopher. 2024. "Opinion | Try Living in Elon Musk's Company Town." Opinion. *The New York Times*, May 24. https://www.nytimes.com/2024/05/24/opinion/elon-musk-spacex-brownsville-texas.html.

Horsey, Julian. 2024. "New MIT Research Proves AGI Was Achieved." *Geeky Gadgets*, November 14. https://www.geeky-gadgets.com/artificial-general-intelligence-advancements/.

Hughes, James. 2004. *Citizen Cyborg: Why Democratic Societies Must Respond to the Redesigned Human of the Future*. Westview. https://doi.org/10.1002/9781118555927.

Hughes, Owen. 2025a. "Just 2 Hours Is All It Takes for AI Agents to Replicate Your Personality with 85% Accuracy." Live Science, January 4. https://www.livescience.com/technology/artificial-intelligence/just-2-hours-is-all-it-takes-for-ai-agents-to-replicate-your-personality-with-85-percent-accuracy.

Huizinga, Johann. [1949] 2000. *Homo Ludens: A Study of the Play-Element in Culture*. Routledge.

Hunt, Katie. 2024. "'A Sort of Superpower': Unexpected Revelations Made Possible by AI in 2024." CNN, December 21. https://www.cnn.com/2024/12/21/science/artificial-intelligence-ai-science-2024.

Huxley, Julian. [1927] 1957. *Religion without Revelation*. Revised edition. New American Library.

——— 1951. "Knowledge, Morality, and Destiny." *Psychiatry: Journal for the Study of Interpersonal Processes* 14 (2): 127–52.

——— 1957. *New Bottles for New Wine*. Harper and Brothers.

——— [1959] 1963. "The Future of Man - Evolutionary Aspects." In *Man and His Future: A Ciba Foundation Volume*, edited by Gordon Wolstenholme. Little, Brown and Company.

Johnson, Rian, dir. 2022. *Glass Onion: A Knives Out Mystery*. Netflix.

Johnson Space Center Office of Communications. 2024. *First Mars Crew Completes Yearlong Simulated Red Planet NASA Mission - NASA*. Crew Health and Performance Exploration Analog (CHAPEA). July 10. https://www.nasa.gov/missions/analog-field-testing/chapea/first-mars-crew-completes-yearlong-simulated-red-planet-nasa-mission/.

Jonze, Spike, dir. 2013. *Her*. Warner Bros.

Joy, Bill. 2000. "Why the Future Doesn't Need Us." *Wired* 8 (4). https://www.wired.com/2000/04/joy-2/.

Kaczynski, Ted. 1996. "Text of Unabomber Manifesto." *New York Times*. https://archive.

nytimes.com/www.nytimes.com/library/national/unabom-manifesto-1.html.

Kaku, Michio. 2018. *The Future of Humanity: Terraforming Mars, Interstellar Travel, Immortality and Our Destiny Beyond Earth*. Allen Lane.

Kass, Leon. 2003. "Letter of Transmittal to the President." *Beyond Therapy: Biotechnology and the Pursuit of Happiness—A Report of the President's Council on Bioethics*. The President's Council on Bioethics.

Katz, Yarden. 2020. *Artificial Whiteness: Politics and Ideology in Artificial Intelligence*. Columbia University Press.

Kaufman, Leeor, and Joe Egender, dirs. 2019. *Unnatural Selection: Cut, Paste, Life*. Documentary. Netflix.

Kepler, Johann. 1967. *Kepler's Somnium: The Dream, Or Posthumous Work on Lunar Astronomy*. Translated by Edward Rosen. University of Wisconsin Press.

Kiesler, Frederick. 1996. "Second Manifesto of Correalism." In *Theories and Documents of Contemporary Art: A Sourcebook of Artists' Writings*, edited by Kristine Stiles and Peter Selz. California Studies in the History of Art 35. University of California Press.

Kim, Junghyung. 2025. "Artificial Intelligence and the Sustainable Future of Co-Creation." *Zygon: Journal of Religion and Science* 60 (1): 18-30. https://doi.org/10.16995/zygon.17283.

Kimmerer, Robin Wall. 2020. *Braiding Sweetgrass: Indigenous Wisdom, Scientific Knowledge and the Teachings of Plants*. Penguin Ecology. Penguin Books.

Kinstler, Linda. 2021. "Opinion | Can Silicon Valley Find God?" Opinion. *The New York Times*, July 16. https://www.nytimes.com/interactive/2021/07/16/opinion/ai-ethics-religion.html.

Kuhn, Thomas. [1962] 1996. *The Structure of Scientific Revolutions*. University of Chicago Press.

Kurzweil, Ray. 1999. *The Age of Spiritual Machines: When Computers Exceed Human Intelligence*. Viking.

——— 2005. *The Singularity Is Near: When Humans Transcend Biology*. Viking.

——— 2024. *The Singularity Is Nearer: When We Merge with AI*. Viking.

Kurzweil, Ray and Terry Grossman. 2005. *Fantastic Voyage: Live Long Enough to Live Forever*. 1. Plume print. Plume.

Lang, Fritz, dir. 1927. *Metropolis*. Parufamet. 153 minutes.

Lanier, Jaron. 2000. "ONE HALF A MANIFESTO." Edge. Accessed March 30, 2025. https://www.edge.org/conversation/jaron_lanier-one-half-a-manifesto.

——— 2010. *You Are Not a Gadget: A Manifesto*. Knopf.

Latour, Bruno. 2009. *Politics of Nature: How to Bring the Sciences into Democracy*. With Catherine Porter. Harvard University Press.

Lawler, Daniel. 2024. "Flood of 'Junk': How AI Is Changing Scientific Publishing." Phys.Org, August 10. https://phys.org/news/2024-08-junk-ai-scientific-publishing.html.

Le Corbusier. 1946. *Towards a New Architecture*. Translated by Frederick Etchells. Praeger.

Ledford, Heidi. 2025. "'Biotech Barbie' Says the Time Has Come to Consider CRISPR Babies. Do Scientists Agree?" *Nature*, ahead of print, November 3. https://doi.org/10.

1038/d41586-025-03554-y.

Leslie, Stuart W. 2017. "Spaces for the Space Age: William Pereira's Aerospace Modernism." In *Blue Sky Metropolis: The Aerospace Century in Southern California*, edited by Peter J. Westwick. Published for Huntington-USC Institute on California and the West by University of California Press and Huntington Library, San Marino, California.

Levy, David. 2006. *Robots Unlimited: Life in a Virtual Age*. A.K. Peters.

Lex Fridman, dir. 2024. *Elon Musk: Neuralink and the Future of Humanity | Lex Fridman Podcast #438*. https://www.youtube.com/watch?v=Kbk9BiPhm7o.

Linehan, Dan. 2008. *SpaceShipOne: An Illustrated History*. With Arthur C. Clarke. Zenith.

MacAskill, William. 2023. *What We Owe the Future: A Million-Year View*. Oneworld.

Mak, Aaron. 2025. "What's up with Peter Thiel and the Antichrist?" POLITICO, October 14. https://www.politico.com/newsletters/digital-future-daily/2025/10/14/whats-up-with-peter-thiel-and-the-antichrist-00608036.

Mancini, Jeannine. 2024. "Larry Page, Google Co-Founder, Said He'd Leave His Fortune To Elon Musk Over Charity Because Of His Plans 'To Go To Mars To Back Up Humanity.'" Benzinga, November 24. https://www.benzinga.com/news/24/11/42154628/larry-page-google-co-founder-said-hed-leave-his-fortune-to-elon-musk-over-charity-because-of-his-plans-to-go-to-mars-to-back-up-humanity.

Marcus, Gary. 2025. "Why My p(Doom) Has Risen, Dramatically." Substack newsletter. *Marcus on AI*, July 15. https://garymarcus.substack.com/p/why-my-pdoom-has-risen-dramatically.

Markoff, John. 2015. *Machines of Loving Grace: The Quest for Common Ground between Humans and Robots*. Ecco, an imprint of HarperCollinsPublishers.

———— 2022. *Whole Earth: The Many Lives of Stewart Brand*. Penguin Press.

Martin, George. 1971. "Brief Proposal on Immortality: An Interim Solution." *Perspectives in Biology and Medicine* 14 (2): 339–40.

Martínez, Sebastián Bruzzone. n.d. "NEOM: Saudi Arabia's Futuristic Mecca." Global Affairs and Strategic Studies. Accessed August 31, 2025. https://en.unav.edu/web/global-affairs/detalle/-/blogs/neom-la-meca-futurista-de-arabia-saudi.

Mason, Christopher E. 2021. *The next 500 Years: Engineering Life to Reach New Worlds*. The MIT Press.

Mayson, Sandra G. 2019. "Bias In, Bias Out." *Yale Law Journal* 128 (8): 2218–300.

McCarthy, John, Marvin L. Minsky, Nathaniel Rochester, and Claude E. Shannon. 2006. "A Proposal for the Dartmouth Summer Research Project on Artificial Intelligence, August 31, 1955." *AI Magazine* 27 (4): 12–14.

McGregor, Philip. 1991. *Rigger Black Book: A Shadowrun Sourcebook*. Fasa.

McKay, Christopher P. 2019. "Prerequisites to Human Activity on Mars: Scientific and Ethical Aspects." *Theology and Science* 17 (3): 317–23. https://doi.org/10.1080/14746700.2019.1633060.

McKinson, Kimberley D. 2021. "Lessons From Mars—and Jamaica—on Sovereignty." *SAPIENS*, September 21. https://www.sapiens.org/culture/accompong/.

McMahon, Liv. 2025. "AI System Resorts to Blackmail If Told It Will Be Removed."

BBC, May 23. https://www.bbc.com/news/articles/cpqeng9d2ogo.

Merrill, Dave and Nadia Lopez. 2023. "Billionaire-Backed Tech Group Says It's Bought All the Land It Needs for Utopian City." *The Seattle Times*, November 4. https://www.seattletimes.com/business/billionaire-backed-tech-group-says-its-bought-all-the-land-it-needs-for-utopian-city/.

Metchnikoff, Élie. 1905. *The Nature of Man: Studies in Optimistic Philosophy*. Translated by P. Chalmers Mitchell. G.P. Putnam's Sons.

Metz, Cade. 2023. "'The Godfather of A.I.' Leaves Google and Warns of Danger Ahead." *The New York Times*, May 1. https://www.nytimes.com/2023/05/01/technology/ai-google-chatbot-engineer-quits-hinton.html.

——— 2024. "Inside OpenAI's Library - The New York Times." *The New York Times*, May 15. https://www.nytimes.com/2024/05/15/technology/openai-library-office.html?smid=nytcore-ios-share&referringSource=articleShare&sgrp=c-cb.

Metz, Cade and Christie Hemm Klok. 2024. "The Old-Fashioned Library at the Heart of the A.I. Boom." *The New York Times*, May 15. https://www.nytimes.com/2024/05/15/technology/openai-library-office.html.

Metzl, Jamie Frederic. 2020. *Hacking Darwin: Genetic Engineering and the Future of Humanity*. Sourcebooks.

Midgley, Mary. [1992] 2002. *Science as Salvation: A Modern Myth and Its Meaning*. Routledge.

Moravec, Hans. 1979. "Today's Computers, Intelligent Machines and Our Future." *Analog*.

——— 1988. *Mind Children: The Future of Robot and Human Intelligence*. Harvard University Press.

——— 1992. "Pigs in Cyberspace." In *Thinking Robots, An Aware Internet, and Cyberpunk Librarians: The 1992 LITA President's Program*, edited by R. Bruce Miller and Milton T. Wolf. Library and Information Technology Association.

——— 1999. *Robot: The Future of Machine and Human Intelligence*. Oxford University Press.

More, Max, and Natasha Vita-More, eds. 2013. *The Transhumanist Reader: Classical and Contemporary Essays on the Science, Technology, and Philosophy of the Human Future*. Wiley-Blackwell.

Moser, Sarah. 2015. "New Cities: Old Wine in New Bottles?" *Dialogues in Human Geography* 5 (1): 31–35. https://doi.org/10.1177/2043820614565867.

Mulkey, Sachi Kitajima. 2024. "Is It Possible to Build a Dream City from Scratch?" Grist, August 1. https://grist.org/cities/california-sustainable-dream-city-from-scratch/.

Musk, Elon. 2017. "Making Humans a Multi-Planetary Species." *New Space* 5 (2): 46–61. https://doi.org/10.1089/space.2017.29009.emu.

——— 2018. "Making Life Multi-Planetary." *New Space* 6 (1): 2–11. https://doi.org/10.1089/space.2018.29013.emu.

Musk, Elon, and Peter Diamandis. 2021. "Elon Musk and Peter Diamandis LIVE on $100M XPRIZE Carbon Removal - YouTube." https://www.youtube.com/watch?v=BN88HPUm6jo&t=745s.

Nagaraj, Nithin. 2020. "AI: From Turing to Sophia." In "Facets of AI" Workshop Hosted by the National Institute of Advanced Studies. Bangalore, India. https://www.youtube.com/watch?v=R4ylT5_Y7Es&feature=youtu.be.

Nagel, Thomas. 1974. "What Is It Like to Be a Bat?" *The Philosophical Review* 83 (4): 435. https://doi.org/10.2307/2183914.

Narayanan, Arvind and Sayash Kapoor. 2025. "AI as Normal Technology." Knight First Amendment Institute, April 15. http://knightcolumbia.org/content/ai-as-normal-technology.

NASA. Undated. "Humans to Mars." NASA. 123. https://www.nasa.gov/humans-in-space/humans-to-mars/.

Natural History, American Museum. 2016. "2016 Isaac Asimov Memorial Debate: Is the Universe a Simulation?" https://www.youtube.com/watch?v=wgSZA3NPpBs.

Négoce, Nicolas, Natasha Booty, and Jonathan Griffin. 2025. "Akon City: Wakanda-Style $6bn Project Abandoned by Senegal." BBC, July 4. https://www.bbc.com/news/articles/c8xvrv21drjo.

Nehru, Jawaharlal. 1946. *The Discovery of India*. Modern Classics. Penguin Books India.

———— 1988. *Jawaharlal Nehru on Science and Society: A Collection of His Writings and Speeches*. Edited by Baldev Singh. Nehru Memorial Museum and Library.

Nelson, Mark. 2018. *Pushing Our Limits: Insights from Biosphere 2*. University of Arizona Press.

NEOM. 2025. "About NEOM: Pioneering the Future of Livability and Business." https://www.neom.com/en-us/about.

Newell, Allen. 1990. "Fairy Tales." In *The Age of Intelligent Machines*, edited by Raymond Kurzweil. The MIT Press.

Newell, Catherine L. 2019. *Destined for the Stars: Faith, the Future, and America's Final Frontier*. University of Pittsburgh Press.

Newton, Casey. 2016. "When Her Best Friend Died, She Used Artificial Intelligence to Keep Talking to Him." TheVerge.Com, October 6. http://www.theverge.com/a/luka-artificial-intelligence-memorial-roman-mazurenko-bot.

Ni, Zhange. 2020. "Reimagining Daoist Alchemy, Decolonizing Transhumanism: The Fantasy of Immortality Cultivation in Twenty-First Century China." *Zygon: Journal of Religion and Science* 55 (3). https://doi.org/10.1111/zygo.12634.

Noble, David. 1999. *The Religion of Technology: The Divinity of Man and the Spirit of Invention*. Penguin.

Novak, Analisa. 2025. "'Godfather of AI' Geoffrey Hinton Warns AI Could Take Control from Humans: 'People Haven't Understood What's Coming.'" CBS News, April 26. https://www.cbsnews.com/news/godfather-of-ai-geoffrey-hinton-ai-warning/.

Nye, David E. 2003. *America as Second Creation: Technology and Narratives of a New Beginning*. MIT University Press.

Oliver, Kendrick. 2013. *To Touch the Face of God: The Sacred, the Profane and the American Space Program, 1957-1975*. New Series in NASA History. Johns Hopkins university press.

O'Neill, Gerard K. 1977. *The High Frontier: Human Colonies in Space*. Morrow.

Orland, Kyle. 2025. "Why Do LLMs Make Stuff up? New Research Peers under the Hood." *Ars Technica*, March 28. https://arstechnica.com/ai/2025/03/why-do-llms-make-stuff-up-new-research-peers-under-the-hood/.

Otto, Rudolph. 1917. *The Idea of the Holy: An Inquiry into the Non-Rational Factor in the Idea of the Divine and Its Relation to the Rational.* Translated by John Harvey. Oxford University Press.

Park, Joon Sung, Carolyn Q. Zou, Aaron Shaw, et al. 2024. "Generative Agent Simulations of 1,000 People." Version 1. Preprint, arXiv. https://doi.org/10.48550/ARXIV.2411.10109.

Parmar, Mayank. 2025. "Researchers Claim ChatGPT O3 Bypassed Shutdown in Controlled Test." BleepingComputer, May 25. https://www.bleepingcomputer.com/news/artificial-intelligence/researchers-claim-chatgpt-o3-bypassed-shutdown-in-controlled-test/.

Paul, Andrew. 2024. "85% of Neuralink Implant Wires Are Already Detached, Says Patient." *Popular Science*, May 21. https://www.popsci.com/health/neuralink-wire-detachment/.

Perrone, Matthew. 2025. "The New Head of the CDC Has No Training in Medicine and Once Helped Peter Thiel Develop Man-Made Islands Floating Outside U.S. Territory." Fortune, August 29. https://fortune.com/2025/08/29/who-is-jim-oneill-acting-head-cdc-peter-thiel-robert-kennedy/.

Prisco, Giulio. 2004. "Engineering Transcendence." http://cosmi2le.com//index.php/site/more/engineering_transcendence.

——— 2013. "Transcendent Engineering." In *The Transhumanist Reader*, edited by Max More and Natasha Vita-More. Wiley-Blackwell.

——— 2021. *Futurist Spaceflight Manifesto*. Independently published.

Project Hyperion. 2025. "Interstellar Generation Ship Design Competition." Project Hyperion. https://www.projecthyperion.org.

Próspera. 2025a. "About Próspera: Innovation in Roatán, Honduras." Próspera. https://www.prospera.co/en/about.

——— 2025b. "Open Application - Próspera." Próspera. https://jobs.prospera.co/27253.

——— 2025c. "Regenerative Medicine Clinic." Próspera City Builders Network. https://community.prospera.co/c/open-opportunities/development-and-deployment-of-next-generation-regenerative-medicine-therapies-in-prospera.

Ptolemy, Barry, dir. 2009. *Transcendent Man*. Docurama.

Rapier, Graham. 2019. "'If You Can't Beth Them Join Them': Elon Musk Says Our Best Hope for Competing with AI Is Becoming Better Cyborgs." *Business Insider*. https://www.businessinsider.com/elon-musk-humans-must-become-cyborgs-to-compete-with-ai-2019-8.

Rebentisch, Hannah, Caroline Thompson, Laurence Côté-Roy, and Sarah Moser. 2020. "Unicorn Planning: Lessons from the Rise and Fall of an American 'Smart' Mega-Development." *Cities* 101 (June): 102686. https://doi.org/10.1016/j.cities.2020.102686.

Redman, Melody, Andrew King, Caroline Watson, and David King. 2016. "What Is CRISPR/Cas9?" *Archives of Disease in Childhood. Education and Practice Edition* 101 (4): 213–15. https://doi.org/10.1136/archdischild-2016-310459.

Rees, Martin, and John McFall. 2024. "Astronomer Royal Lord Rees of Ludlow: Lord Speaker." UK Parliament's House of Lords, March 26. https://www.parliament.uk/

business/lords/house-of-lords-podcast/astronomer-royal-lord-rees-of-ludlow-lord-speakers-corner/.

Regis, Ed. 1994. "Meet the Extropians." *Wired*, October 1. Accessed July 10, 2025. https://www.wired.com/1994/10/extropians/.

Reilly, Patrick. 2023. "Tech Mogul Bryan Johnson Undergoes Shock Therapy on Penis to Get 'the Erections of an 18 Year Old.'" *NY Post*, November 9. https://nypost.com/2023/11/09/health/tech-mogul-bryan-johnson-undergoes-shock-therapy-on-penis-to-get-the-erections-of-an-18-year-old/.

Robison, Kylie. 2025. "Claude Fans Threw a Funeral for Anthropic's Retired AI Model." *Wired*, August 5. https://www.wired.com/story/claude-3-sonnet-funeral-san-francisco/.

Rollett, Charles. 2025. "Startup Co-Founded by Longevity Guru Peter Attia Emerges from Stealth." *TechCrunch*, February 28. https://techcrunch.com/2025/02/28/startup-co-founded-by-longevity-guru-peter-attia-emerges-from-stealth/.

Ropek, Lucas. 2023. "Shadowy Tech Goons Want to Build a New City in California. What Could Go Wrong?" *Gizmodo*, August 28. https://gizmodo.com/michael-moritz-reid-hoffman-flannery-associates-land-bu-1850780238.

——— 2025. "Longevity-Obsessed Tech Millionaire Discontinues De-Aging Drug Out of Concerns That It Aged Him." *Gizmodo*, January 13. https://gizmodo.com/longevity-obsessed-tech-millionaire-discontinues-de-aging-drug-out-of-concerns-that-it-aged-him-2000549377.

Rubenstein, Mary-Jane. 2022. *Astrotopia: The Dangerous Religion of the Corporate Space Race*. The University of Chicago Press.

Russo, Anthony and Joe Russo, dirs. 2025. *The Electric State*. Netflix.

Schaffer, Simon. 2002. "The Devices of Iconoclasm." In *Iconoclash: Beyond the Image Wars in Science, Religion, and Art*, edited by Bruno Latour and Peter Weibel. MIT University Press and ZKM Center for Art and Media.

Schott, Gareth. 2010. "Dawn of the Digital Dead: The Zombie as Interactive Social Satire in American Popular Culture." *Australasian Journal of American Studies* 29 (1): 61–75.

Science Gallery Bengaluru, dir. 2022. *Can Machines Come Alive?* https://www.youtube.com/watch?v=U4fjHeirVtM.

Scott, Ridley, dir. 1982. *Blade Runner*. Warner Bros. 117 minutes.

Secretary Doug Burgum [@SecretaryBurgum]. 2025. "The mission of the U.S. Fish and Wildlife Service is to work with others to 'conserve, protect, and enhance fish, wildlife, plants, and their habitats for the continuing benefit of the American people.' The Department of the Interior is excited about the potential of." Tweet. Twitter, April 7. https://x.com/SecretaryBurgum/status/1909345951069651032.

Segalov, Michael. 2024. "'With Brain Preservation, Nobody Has to Die': Meet the Neuroscientist Who Believes Life Could Be Eternal." *The Guardian*, December 1. https://www.theguardian.com/science/2024/dec/01/with-brain-preservation-nobody-has-to-die-meet-the-neuroscientist-who-believes-life-could-be-eternal.

Sharma, Neha Tandon. 2024. "Saudi Crown Prince MBS Is Planning to Revolutionize Construction at His $1.5 Trillion Neom Megaproject by Using Cutting-Edge

Robotics to Replace on-Site Workers." Luxurylaunches, December 13. https://luxu
rylaunches.com/other_stuff/neom-robotics-12132024.php.

Shaw, George. 1921. *Back to Methuselah. A Metabiological Pentateuch*. Brentano.

Shlapentokh, Dmitry. 1996. "Bolshevism as a Fedorovian regime*." *Cahiers du monde russe* 37 (4): 429–65. https://doi.org/10.3406/cmr.1996.2473.

Singler, Beth. 2025. *Religion and Artificial Intelligence: An Introduction*. Routledge.

Smith, Jonathan Z. 1982. *Imagining Religion: From Babylon to Jonestown*. University of Chicago Press.

Smith, Kiona N. 2018. "The Correction Heard 'Round The World: When The New York Times Apologized to Robert Goddard." *Forbes*, July 19. https://www.forbes.com/sites/kionasmith/2018/07/19/the-correction-heard-round-the-world-when-the-new-york-times-apologized-to-robert-goddard/.

Solano Together. 2025. "The East Solano Plan." Solano Together. https://www.solanotogether.org/eastsolanoplan.

Song, Yong Sup. 2021. "Religious AI as an Option to the Risks of Superintelligence: A Protestant Theological Perspective." *Theology and Science* 19 (1): 65–78. https://doi.org/10.1080/14746700.2020.1825196.

Song, Yong Sup, and Robert M. Geraci. 2023. "Marginalization and Transcendence in Transhumanism and Minjung Theology." *Zygon: Journal of Religion and Science* 58 (1). https://doi.org/10.1111/zygo.12834.

——— 2025. "Cultural Theology and Social Robotics: The Intersection of Religion and Science in Ethical AI." *Religion* 55 (3): 715–37. https://doi.org/10.1080/0048721X.2025.2502294.

Stark, Rodney and William Sims Bainbridge. 1985. *The Future of Religion: Secularization, Revival, and Cult Formation*. University of California Press.

Stephenson, Neal. 1992. *Snow Crash*. Bantam.

——— 2015. *Seveneves*. William Morrow.

Stock, Gregory. 2003. *Redesigning Humans: Our Inevitable Genetic Future*. 1. Mariner Books ed. A Mariner Book. Houghton Mifflin.

Strathmore, Lucy. 2025. "JD Vance Uses His Mom to Deflect Billionaire Donor Question in Interview." https://2paragraphs.com/2025/01/jd-vance-uses-his-mom-to-deflect-billionaire-donor-question-in-interview/.

Subramanian, Samanth. 2019. *A Dominant Character: The Radical Science and Restless Politics of J.B.S. Haldane*. Simon & Schuster.

Svyatogor, Alexander. 2018. "Biocosmist Poetics." In *Russian Cosmism*, edited by Boris Groys. Eflux and The MIT Press.

Swift, David. 1990. *SETI Pioneers: Scientists Talk about Their Search for Extraterrestrial Intelligence*. With Frank Drake. University of Arizona Press.

Swisher, Kara. 2020. "Elon Musk: 'A.I. Doesn't Need to Hate Us to Destroy Us.'" *The New York Times*, September 28. https://www.nytimes.com/2020/09/28/opinion/sway-kara-swisher-elon-musk.html.

Tangermann, Victor. 2024. "Elon Musk's Plans for a City on Mars Will Likely End in Horrifying Mass Death." *Futurism*, November 30. https://futurism.com/elon-musk-city-mars-death-catastrophe.

Tarantola, Andrew. 2023. "Offworld 'company Towns' Are the Wrong Way to Settle the Solar System." *Engadget*, December 10. https://www.engadget.com/hitting-the-books-a-city-on-mars-kelly-and-zach-weinersmith-penguin-153023805.html.

Tegmark, Max. 2017. *Life 3.0: Being Human in the Age of Artificial Intelligence*. Knopf.

Teilhard de Chardin, Pierre. 1955. *The Phenomenon of Man*. Translated by Bernard Wall. Harper & Brothers.

Telosa. n.d. "Telosa | City of the Future." Telosa. Accessed August 31, 2025. https://cityoftelosa.com/.

The Wachowskis, dir. 1999. *The Matrix*. Warner Bros.

Theta Noir. 2025. "Myth — THETA NOIR | Wake Up To The Collective Mind: MENA." Theta Noir. https://thetanoir.com/Myth.

Tingley, Brett. 2024. "An Asteroid Hit Earth Just Hours after Being Detected. It Was the 3rd 'imminent Impactor' of 2024." Yahoo News, November 8. https://www.yahoo.com/news/asteroid-hit-earth-just-hours-203055120.html.

Tirosh-Samuelson, Hava. 2013. "Wrestling with Transhumanism." In *The Transhumanist Reader: Classical and Contemporary Essays on the Science, Technology, and Philosophy of the Human Future*, edited by Max More and Natasha Vita-More. Wiley-Blackwell.

——— 2018. "In Pursuit of Perfection: The Misguided Transhumanist Vision." *Theology and Science* 16 (2): 200–222.

Tremlin, Todd. 2006. *Minds and Gods: The Cognitive Foundations of Religion*. Oxford University Press.

Turing, A. M. 1950. "Computing Machinery and Intelligence." *Mind* LIX (236): 433–60. https://doi.org/10.1093/mind/LIX.236.433.

Tyson, Mark. 2025. "Latest OpenAI Models 'Sabotaged a Shutdown Mechanism' despite Commands to the Contrary." Tom's Hardware, May 26. https://www.tomshardware.com/tech-industry/artificial-intelligence/latest-openai-models-sabotaged-a-shutdown-mechanism-despite-commands-to-the-contrary.

Ulam, Stanislaw. 1958. "Tribute to John von Neumann." *Bulletin of the American Mathematical Society* 64 (3, pt. 2): 1–49.

Vance, Ashlee. 2012. "The Ray Kurzweil Show, Now at the Googleplex." *Business Insider* 4310: 55–56.

Verne, Jules. [1865] 2019. *From the Earth to the Moon: 100th Anniversary Collection*. Translated by Louis Mercier. Seawolf Press.

Villarreal, Mireya, Jim Scholz, and Joe LoCascio. 2025. "How Elon Musk's SpaceX May Take over a Corner of the Texas Coast." ABC News. Accessed May 3, 2025. https://abcnews.go.com/Politics/elon-musks-spacex-corner-texas-coast/story?id=121361452.

Vinge, Vernor. 1981. "True Names." In *True Names and the Opening of the Cyberspace Frontier*, edited by James Frenkel. Tor Books.

——— 1993. "Technological Singularity." https://www.frc.ri.cmu.edu/~hpm/book98/com.ch1/vinge.singularity.html.

——— 1983. "First Word." *OMNI* 5 (1): 5.

Wachhorst, Wyn. 2001. *The Dream of Spaceflight: Essays on the Near Edge of Infinity*. Da Capo Press.

Wadhawan, Vinod Kumar. 2007a. "Robots of the Future." *Resonance* 12 (7): 61–78.

———— 2007b. *Smart Structures: Blurring the Distinction between the Living and the Nonliving*. Oxford University Press.

Wahrman, Miryam Z. 2002. *Brave New Judaism: When Science and Scripture Collide*. Brandeis University Press.

Wakefield, Jane. 2019. "Elon Musk Reveals Brain-Hacking Plans." *BBC News*. https://www.bbc.com/news/technology-49004004.

Warhol, Andy. 1996. "Warhol in His Own Words." In *Theories and Documents of Contemporary Art: A Sourcebook of Artists' Writings*, edited by Kristine Stiles and Peter Howard Selz. California Studies in the History of Art 35. University of California Press.

Warwick, Kevin. 1997. *March of the Machines: The Breakthrough in Artificial Intelligence*. University of Illinois Press.

———— 2002. *I, Cyborg*. Century.

———— 2003. "Cyborg Morals, Cyborg Values, Cyborg Ethics." *Ethics and Information Technology* 5 (3): 131–37.

Weber, Tomas. 2024. "The Inside Story behind Decoding an Ancient Herculaneum Scroll." Scientific American, April 1. https://www.scientificamerican.com/article/inside-the-ai-competition-that-decoded-an-ancient-scroll-and-changed/.

Weindling, Paul. 2012. "'Julian Huxley and the Continuity of Eugenics in Twentieth-Century Britain.'" *Journal of Modern European History = Zeitschrift Fur Moderne Europaische Geschichte = Revue d'histoire Europeenne Contemporaine* 10 (4): 480–99. https://doi.org/10.17104/1611-8944_2012_4.

Weizenbaum, Joseph. 1976. *Computer Power and Human Reason: From Judgment to Calculation*. W.H. Freeman and Company.

Whipple, Tom. 2024. "Should You Freeze Your Brain? These Scientists Think It's Worth a Shot." *The Times*, November 16. https://www.thetimes.com/uk/science/article/should-you-freeze-your-brain-these-scientists-think-its-worth-a-shot-2tlr6sd7r.

White, Andrew D. 1896. *A History of the Warfare of Science with Theology in Christendom*. Vol. 1. D. Appleton and Company.

White, David Gordon. 2011. *Sinister Yogis*. University of Chicago Press.

Wiener, Norbert. [1950] 1954. *The Human Use of Human Beings: Cybernetics and Society*. Rev. Ed. Doubleday Anchor.

———— 1964. *God & Golem, Inc.: A Comment on Certain Points Where Cybernetics Impinges on Religion*. The M.I.T. Press.

Wikipedia. 2025. "Biosphere 2." In *Wikipedia*. July 30. https://en.wikipedia.org/w/index.php?title=Biosphere_2&oldid=1303270406.

Wright, Oliver and Mark Sellman. 2024. "Britain Must Treat Tech Giants like Nation States, Minister Warns." *The Times* (UK), November 12. https://www.thetimes.com/uk/politics/article/britain-must-treat-tech-giants-like-nation-states-minister-warns-ktmm5vmc9.

Wright, Webb. 2025. "Spiritual Influencers Say 'Sentient' AI Can Help You Solve Life's Mysteries." *Wired*, September 2. https://www.wired.com/story/spiritual-influencers-say-sentient-ai-can-help-you-solve-lifes-mysteries/.

XPRIZE, dir. 2021. *Elon Musk and Peter Diamandis LIVE on $100M XPRIZE Carbon Removal.* https://www.youtube.com/watch?v=BN88HPUm6jo.

Yaron, Barr. 2023. "State of AI Engineering 2023." Elemental-Croissant-32a on Notion (Amplify Partners). https://elemental-croissant-32a.notion.site/State-of-AI-Engineering-2023-20c09dc1767f45988ee1f479b4a84135#694f89e8691148cb855220ec05e9c631.

Yogananda, Paramahansa. 1946. *Autobiography of a Yogi.* Yogoda Satsanga Society of India.

Young, George. 2012. *The Russian Cosmists: The Esoteric Futurism of Nikolai Fedorov and His Followers.* Oxford University Press.

Yudkowsky, Eliezer. 2001. *Creating Friendly AI 1.0: The Analysis and Design of Benevolent Goal Architectures.* The Singularity Institute.

Zahn, Max. 2024. "'I Love You Robo-Dad': Meet a Family Using AI to Preserve Loved Ones after Death." ABC News, July 24. https://abcnews.go.com/Business/love-robo-dad-meet-family-ai-preserve-loved/story?id=111756468.

Zeleznikow-Johnston, Ariel, Emil F. Kendziorra, and Andrew Thomas McKenzie. 2024. "What Are Memories Made of? A Survey of Neuroscientists on the Structural Basis of Long-Term Memory." Preprint, November 7. https://doi.org/10.31234/osf.io/keq7w.

Zuboff, Shoshana. 2015. "Big Other: Surveillance Capitalism and the Prospects of an Information Civilization." *Journal of Information Technology* 30 (1): 75–89.

Zubrin, Robert. 2019. "Why We Earthlings Should Colonize Mars!" *Theology and Science* 17 (3): 305–16. https://doi.org/10.1080/14746700.2019.1632519.

ACKNOWLEDGMENTS

There are always more people who deserve credit than my limited brain can accommodate: but it's always easy to remember Jovi, Zion, and Dorian. They support me through every professional and personal enterprise, including the ones (like this book) that become far more consuming than I expected.

There are many people who offered counsel when I needed it. Almost two decades ago, Matt Mason opened the doors of Carnegie Mellon University's famed Robotics Institute to me, and my time there has influenced me in a dozen different directions. Aaron Ulrey, Rodney Sebastian, Vince Biondo, and Tom Ferguson shared their insight into angels and saints in various traditions. Jovi Geraci weighed in on American religious history. Cat Newell did so on the context of American spaceflight. Anya Bernstein was a wonderful resource on Pleistocene Park, even generously sharing her book in progress. Over the years, John Markoff has been a great email partner about Silicon Valley and deserves special thanks for so often championing my work, even putting his own reputation on the line for me on more than one occasion.

In academia, I often say, we should be having fun. The learning, the talking, the sharing. It should be fun! And there have been many folks who've shared my own joy with me while I worked on this book. There were my closest friends at my previous academic home (especially Cory, Kevin, and Steve). And there are friends at my new academic home (especially Scott, Dan, and Anne). These kinds of colleagues make our job so much more than it otherwise would be. Outside my institutional homes, some conversation partners have been especially committed over the years. Bill Bainbridge has been there for professional conversations and connections too numerous to detail. Alex Ornella is always a partner in thinking about technology and has become a member of our family. Yong Sup Song is not just a collaborator, but a brother – he has done far more for me than I could possibly deserve.

Sprinkled throughout this book are references to old pen and paper roleplaying games. Those games were essential to my formation as a person and as a scholar. As I've said before and will say again, they taught me to dream with rigor. So, not only am I indebted to the games' innovators and ongoing contributors, but I am grateful to the folks who brought them into my life. My brother, Vinny, gave me my first *Dungeons & Dragons* books; without him, my life would have been much different! My brother, Cody, played with me from our youth to our adulthood and also introduced me to *Shadowrun*. Our countless conversations about myth, technology, and everything in between inform most of what I've done as a scholar.

Many people have given me chances to speak and work out the ideas that became this book. The Knox College Board of Trustees invited me to their meeting in the fall of 2025 as a new faculty member and listened while I roughed out the futureproofing religion. Beth Singler and the folks at the University of Zurich invited me to share some thoughts at a keynote lecture in 2025. A shocking number of attentive and wonderful audiences have greeted me in Korea these past few years. I've had the amazing good fortune to present to the students and faculty of Methodist University, Presbyterian University and Seminary, Wonkwang University, Seoul Christian University, Yonsei University, Kyung Hee University, Youngnam Theological University and Seminary, Sungkyul University, Sogang University, the Society for Cultural Theology, and the Human Technology Symbiosis Network. Some of those universities even invited me back more than once over the years I have visited. If that's not a sign of wonderful hospitality, I don't know what is.

A few folks did heavy lifting in bringing this book together. John, Matt, and Ilia Delio were generous enough to read a rough draft of this book in advance, and even to say kind things about it. My students at Manhattan College and Knox College have let me bounce many of these ideas off of them, and have, I think, had fun in class with me. I had fun, anyway. I was unbelievably pleased to work with Richard Borge on the amazing cover art and the animated version that graces my social media outreach. Skye Kilaen helped me with the intricacies of self-publication. It was her effort and skill which opened the door for me to publish a book at a reasonable price. Through the Knight Fund for the Study of Religion and Culture, Knox College provided financial support that defrayed some of the publishing costs – and I am grateful for each and every dollar!

Without Mary Kent Knight there would be no Knight Fund. And without Mary Kent Knight, Knox would be greatly impoverished. I thank her for her brilliance, her good cheer, and her commitment to a better world.

Anyone who teaches at a small liberal arts college knows there are remarkable opportunities to help and be helped by one's students. I thought about dedicating this book to them, but the students closest to my heart took primacy instead. Nevertheless, I want to recognize the folks who've often made teaching joyful and special, beginning with the ones who were at my previous institution as I created and three times taught the class where these ideas began. Beyond them, I humbly thank the many students who have been a close part of my intellectual and personal journey, especially: Page, Jesse, Vanessa, Sammy, Stan, Katie, Gail, Catie, Jon, Doug, Kate, Bridget, Nat, Sam, Victoria, Grace, Zand, Elle, Amy, Jake, Chris, Jake, Walter, Ysa, Jesus, Kem, Tyler, Joy, and Mao. Working with you and knowing you has been a delight.

Poppy deserves a special thank you because sometimes – when I was obviously supposed to be playing with her or feeding her snacks – I instead worked on this book. She is truly gracious.

And for the most gracious of all I return to my family. Jovi and my children are the start and finish of my day, the beginning and end of my life. I love them in all the clichéd ways, and in many more ways than that.

ABOUT THE AUTHOR

Robert M Geraci is Knight Distinguished Chair for the Study of Religion & Culture at Knox College in Galesburg, IL. He has been a visiting researcher at Carnegie Mellon University's Robotics Institute, the Indian Institute of Science, and the National Institute for Advanced Studies in Bangalore, India. His research has been supported by the US National Science Foundation, the Republic of Korea National Research Foundation, the American Academy of Religion, and two Fulbright-Nehru research awards.

Geraci has lived in India, collaborates in Korea, and studies the world of science and technology with a lens grounded in religious thought and practice. His projects engage technologies from handloom weaving to virtual worlds, but most of his research has to do with artificial intelligence. He has been reflecting on the religious narratives told by AI researchers for more than two decades. He enjoys hiking, kayaking, and *Dungeons & Dragons*.

His website is at https://robertgeraci.com.

Other delightful books by Robert M Geraci

- *Apocalyptic AI: Visions of Heaven in Robotics, Artificial Intelligence, and Virtual Reality* (Oxford 2010)
- *Virtually Sacred: Myth and Meaning in World of Warcraft and Second Life* (Oxford 2014)
- *Temples of Modernity: Nationalism, Hinduism, and Transhumanism in South Indian Science* (Lexington 2018)
- *Futures of Artificial Intelligence: Perspectives from India and the U.S.* (Oxford 2022)
- *re:Generated Prairie* (limited edition artbook, in collaboration with Michael Takeo Magruder for the artist's *re:Generated Prairie* exhibition at Knox College, 2025)